HANDBOOK FOR SCIENCE PUBLIC INFORMATION OFFICERS

What Editors Want
Philippa J. Benson and Susan C. Silver

Permissions, A Survival Guide
Susan M. Bielstein

The Craft of Research
Wayne C. Booth, Gregory G. Colomb, and Joseph M. Williams

Writing Science in Plain English
Anne E. Greene

The Craft of Scientific Communication
Joseph E. Harmon and Alan G. Gross

Storycraft
Jack Hart

The Chicago Guide to Writing about Multivariate Analysis
Jane E. Miller

The Chicago Guide to Writing about Numbers
Jane E. Miller

The Chicago Guide to Communicating Science
Scott L. Montgomery

A Handbook of Biological Illustration
Frances W. Zweifel

Handbook for **Science Public Information Officers**

W. MATTHEW SHIPMAN

THE UNIVERSITY OF CHICAGO PRESS

Chicago and London

W. MATTHEW SHIPMAN is a science writer and public information officer at North Carolina State University.

The University of Chicago Press, Chicago 60637
The University of Chicago Press, Ltd., London

Printed in the United States of America

24 23 22 21 20 19 18 17 16 15 1 2 3 4 5

ISBN-13: 978-0-226-17932-2 (cloth)
ISBN-13: 978-0-226-17946-9 (paper)
ISBN-13: 978-0-226-17963-6 (e-book)
DOI: 10.7208/chicago/9780226179636.001.0001

Library of Congress Cataloging-in-Publication Data

Shipman, W. Matthew, author.
Handbook for science public information officers / W. Matthew Shipman.
pages; cm. — (Chicago guides to writing, editing, and publishing)
Includes index.
ISBN 978-0-226-17932-2 (cloth: alk. paper) — ISBN 0-226-17932-X (cloth: alk. paper) — ISBN 978-0-226-17946-9 (pbk.: alk. paper) — ISBN 0-226-17946-X (pbk.: alk. paper) — ISBN 978-0-226-17963-6 (e-book) 1. Communication in science—Handbooks, manuals, etc. 2. Technical writing—Handbooks, manuals, etc. 3. Communication of technical information—Handbooks, manuals, etc. 4. Science news. I. Title. II. Series: Chicago guides to writing, editing, and publishing.
Q223.S45 2015
808.06'65—dc23 2014040938

♾ This paper meets the requirements of ANSI/NISO Z39.48-1992 (Permanence of Paper).

CONTENTS

INTRODUCTION
Why Science PIOs Matter (and Communication 101)

Research institutions come in all shapes and sizes, from federal agencies and government laboratories to universities and nonprofit research centers. They are home to researchers engaged in furthering basic science, engineers developing new technologies, and scientists seeking to cure diseases. These institutions share a drive to answer questions and advance the state of human knowledge. They also share a desire to communicate with the public—a catch-all term that we use to refer to a wide variety of audiences who are often unable to understand the technical jargon found in scientific papers.

Research institutions need to communicate for a variety of reasons, including to secure funding from state or federal legislators, to attract researchers or graduate students, and to disseminate research findings in order to comply with the terms of a grant agreement. To accomplish any one of these goals, the institution will need to reach a variety of key audiences: legislators, federal agency staff, graduate students in a specific discipline, and the so-called "general public," whose opinion can sway the decisions of public officials of all stripes. But scientists and research administrators do not always have the time or skills to promote their research findings to those audiences. This is where a public information officer (PIO) comes in.

A PIO's job is to make his or her employer look good. A science PIO's job is to make his or her research institution look good by emphasizing the importance of the institution's work. Why is the institution's work important? How does the work advance science or medicine? The goal of this book is to help science PIOs promote their employers' science and technology research, regardless of whether the PIO works for a university, a government laboratory, a nonprofit organization, or any other research-driven institution.

The landscape of science news has shifted in recent decades. While 95 U.S. newspapers had sections devoted to science coverage in 1989, there were only 31 papers that had sections devoted to science or health in 2009.[1] As a result of changes like this, it seems that there are now more science PIOs than science reporters. This makes the role of science PIOs both more difficult and more important: more difficult because PIOs are competing for the attention of fewer reporters; more important because many news sources are relying on content generated by PIOs, bypassing reporters altogether. PIOs will not, and should not, ever replace reporters. But it is increasingly important for PIOs to be responsible, effective science communicators who can explain complex science to non-expert audiences.

But being a good science communicator is only one part of the job. Among other things, science PIOs also need to know how to pitch stories to reporters without being annoying, give scientists the skills they need to navigate a news interview, use social media effectively, deal with really bad news about their researchers or institutions, and be able to measure if what they're doing is making a difference to the institution's work.

This book is not designed to tackle all aspects of institutional communication (e.g., it won't address how to notify employees about human resources issues). Instead, it focuses on issues related to communicating about science and should be useful regardless of your level of experience. For students or new science PIOs, it should serve as an introduction that provides smart, practical advice you can use immediately. This book is not designed to turn you into an instant expert on every aspect of media relations and science outreach—*that* would be a much longer book. Instead, my goal is to give you the tools that you need to make good decisions and serve as an effective liaison between your research institution and the public, with a focus on media relations and working with reporters. In

1. Cristine Russell, "Covering Controversial Science: Improving Reporting on Science and Public Policy," in *Science and the Media*, ed. Donald Kennedy and Geneva Overholser (American Academy of Arts and Sciences, 2010), 18–19.

short, I want to teach you how to think critically about what science PIOs do and how they do it. For established science PIOs, I hope this book will remind you of the basics, offer some tips you haven't thought of, and possibly challenge a few preconceived notions. (I'm already looking forward to the feedback I'll get from readers.)

Before we dive in to chapter 1, I want to offer a brief overview of communication basics. These ideas are essential to any communication effort, and some of the components (audience, goals, etc.) crop up throughout the book.

KNOW YOUR AUDIENCE

Whom are you trying to reach? This affects everything else you do. For example, if you're trying to reach an audience of biochemists, you can write in technical language that might baffle anyone else. If you want to reach scientists in other fields, you'll have to tone down the biochemistry jargon, but you can assume familiarity with the scientific process. And if you want to reach, say, politicians, you can assume nothing. I often think of my audience as "the intelligent non-expert." This enables me to reach a fairly broad audience, including all of the groups I just mentioned. Note: If there is any way to avoid it, don't say you're trying to reach the "general public"; it's an incredibly imprecise term. The more specific you are when defining your target audiences, the easier it will be to determine if you're reaching them.

SET GOALS

Defining your audience and setting goals are intertwined. It boils down to a fundamental question: What do you want your target audience to do? It could be as specific as asking registered voters to contact Congress about their support for NASA funding, or as general as helping people to understand the importance of biodiversity. Maybe you simply want to disseminate research findings that you think are interesting in an effort to get people excited about science.

As a science PIO at a public university, I usually have multiple goals for anything I write. Among other things, I want to raise the profile of the institution in a positive way by highlighting faculty research; draw attention to federally funded research (which makes federal funding agencies happy); attract potential students and faculty to our research programs; reach researchers at other institutions who may be interested in formal or informal collaboration with our faculty; and make our students and alumni feel good about the work being done at their university and alma mater.

Knowing your goals helps you better define your key audiences. It's a simple question: Which audiences do you need to reach in order to help your institution make progress toward its goals? This affects the stories you write, how you write them, and where you pitch them. The more specific your goals are, the easier it will be to tell if you're reaching them.

METRICS

You need to have some way of measuring whether you're making progress toward your goals. I've devoted an entire chapter to this topic, so I won't go into an enormous amount of detail here. Metrics let you know how well (or how poorly) your communication efforts are working. The closer you can tie your metrics to your goals, the better off you'll be. For example, if your goal is to recruit people to participate in a citizen science study at your institution, your metric should be the number of people who have signed up—not the number of people who read a blog post you wrote about it.

EVALUATE AND ADJUST

Once you have established metrics, you need to use them. Are your communication efforts working? If not, what should you do about it? Let's stick with the citizen science idea. If you wanted to enlist 750 participants, and you've enrolled 500 in less than a week, keep doing what you're doing. But if you've only signed up 36 participants, you may want to reevaluate your communication approach. If you were relying solely on social media to drive traffic to your citizen science project, you need to branch out. Maybe try contacting reporters in your target markets, to see if they might be interested in doing a story on your project. As FDR once said (in a completely different context), "Take a method and try it. If it fails, admit it frankly and try another. But above all, try something."

Hopefully, this book will help you try something that works.

FINDING STORIES AND DECIDING WHAT TO WRITE ABOUT

Science PIOs are tasked with making complex scientific research accessible to non-expert audiences, but there is more to the job than writing good science stories. You also have to find the right science stories to tell. Perhaps the biggest challenge facing any science PIO is discovering interesting, important, and timely science stories. Threshold questions that you should always ask about a potential story are "What is this, and why would anyone care about it?" If you can answer those questions, you're off to a good start. Timeliness is critical; it is difficult to interest reporters or the public in discoveries or journal articles that are months or years old.

This chapter will focus on issues such as building relationships with your institution's researchers and how to determine what research findings are worth sharing. It will also tackle issues such as how and whether to promote research-related news, such as grant awards.

USE GOALS TO PICK STORIES

Before you can find a good story, you need to know how your institution views itself. What are your institution's goals and priorities? What sort of image does it want to project? You need to know what is important to your institution in order to identify stories that will help the institution make progress toward its goals. Does your employer want to be seen as an innovator? Viewed as a practical problem-solver? Or thought of as a bastion of basic science? Does your institution want to shore up support for continued (or increased) state or federal funding? Is it trying to recruit new researchers in a specific field?

Understanding how your institution views itself is important because you'll need to reach the right audiences if you want to help your institution achieve its goals. And you'll need to find the right stories to reach those audiences. For example, if your institution is trying to create partnerships with private-sector businesses, you should write about research that has obvious market applications. If your institution is trying to attract new faculty in materials science, you'll want to highlight the new research facilities that give materials science researchers access to state-of-the-art equipment. It's worth noting that communication plans are a great tool for focusing your audience identification efforts and developing key messages for those audiences. (Communication plans are discussed at length in chapter 8.)

You can begin to learn about your institution's priorities by reading its mission statement and strategic plan (if they exist). Talk to your boss. Odds are that your employer has a document laying out its communication goals; if it does, you need to familiarize yourself with those goals. Once you've done your homework, talk to management. This will help you get to know your organization from the top down. After all, if the institution has hired a PIO, someone in a position of authority thinks it is important to communicate the institution's research to the public.

Talking to management (e.g., in a university, this includes deans or department heads) will help you understand how an organization works. It will also give you a better understanding of the institution's goals, which will help you determine which stories they think are most important. If the organization is a large one, management can also tell you which researchers are the most productive, helping you prioritize which researchers you want to meet with first. Also, some researchers will be more willing to meet with you if you say that their boss referred you to them. And the researchers are in the best position to tell you about great stories.

KNOW WHOM YOU'RE WORKING WITH

If you are trying to keep track of the research at your institution, you need to build relationships with the researchers themselves. They are the only people who know exactly what's going on in the lab and who are tracking the progress of their publications, which give you an opportunity to share their stories.

Follow the News

A science PIO needs to follow the news on a regular basis. If you're familiar with stories that are in the public eye, you can identify opportunities to capitalize on the expertise of your institution and its researchers. Current news can also help you find a good angle for promoting your institution's research, making it more likely that you'll catch the attention of reporters who cover the relevant field of study.

Most researchers are interested in sharing their findings with their peers. However, many researchers have no desire to talk to reporters about their work. By extension, this means that many researchers have little interest in working with their PIOs. Try to work with them anyway. Once you've met with management, make an effort to meet with as many researchers as possible. Go to their offices and tour their labs. Learn about what they are working on and why. If you can find out what they are passionate about and what drew them to their research in the first place, you will begin to understand what sort of stories you can tell about their work.

The most important thing you can do when meeting with researchers is to give them an understanding of who you are, where you fit into the organization, and how you can work with them, now and in the future.

SELFISH REASONS FOR SCIENTISTS TO PROMOTE THEIR FINDINGS

One of the first things I ask researchers when we meet is whether they think it is important to promote their work outside of conferences and peer-reviewed journals. Sometimes they say yes, and I can skip straight to the part of the conversation where I explain how we can work together. But researchers are often unsure of why they should promote their work and are only meeting with me out of a sense of obligation to their employer. When I sense that researchers are ambivalent about working with me, I highlight some purely selfish reasons for them to publicize their findings. I focus on selfish reasons because scientists are only human. In theory, I could talk about the need to inspire the next generation of scientists or to address the paltry state of science literacy, but if I'm going to ask people to take time away from other obligations, I have to do better than that. I need to explain what's in it for them.

So I have come up with a list of reasons why a self-interested scientist should promote his or her work. To be clear, all of these reasons refer to benefits that I have personally witnessed. While all of these things don't happen every time someone makes an effort to promote his or her work, they are not purely hypothetical, either. And these are not simply points that you should make when trying to convince a researcher to work with you—they are details that you should bear in mind when choosing, writing, and promoting research stories. The following reasons for promoting their work will benefit researchers but will also help to boost the prestige of your institution.

Citations

Researchers care about citation rates. They want other scientists to read their journal articles and cite them in their papers. This is an important metric for determining success in the research community. Research shows that news coverage of a journal article significantly increases the number of times it is cited,[1] and promoting one's research is almost the only way to get any news coverage. In short, if scientists promote their publications, more people are likely to see the papers and the number of citations will likely go up.

Informal Networking

I know of hundreds of examples of scientists who have been contacted by researchers at other institutions after publicizing their work. Sometimes it's just a pat on the back, which is nice. But sometimes there are tangible benefits, such as proposals to share data that will advance the efforts of all parties involved. These commu-

1. David P. Phillips et al., "Importance of the Lay Press in the Transmission of Medical Knowledge to the Scientific Community," *New England Journal of Medicine* 325 (1991): 1180–83.

nications can also include a lot of questions and sharing of ideas, which can lead to the next potential benefit.

Creating Opportunities for Formal Collaborations

I can think of several instances where publicizing the findings of one research project has led to an invitation for that researcher to be part of a new or emerging research project. Often these invitations take the form of inter-institutional or interdisciplinary grant proposals—and words like "inter-institutional" and "interdisciplinary" are increasingly popular with the folks who review grant proposals. Research funding is thin on the ground, and opportunities to participate in viable grant proposals are valuable.

Making Funding Agencies Happy

There's a reason why most grant proposals include a section asking how researchers plan to disseminate their findings. In the United States, most federally funded agencies want the public to know about the work they are supporting. It helps give agencies the political support they need to get additional funding in the future.

Governmental interest in science communication is not limited to the United States, either. From the European Union to Brazil, government agencies, ministries, associations, and foundations highlight the importance of science communication. The languages differ, but the goals are similar. In New Zealand, for example, the Ministry of Research, Science and Technology cites "engaging New Zealanders with science and technology" as a key objective. Meanwhile, Chinese law tasks the Chinese Association for Science and Technology with disseminating "scientific and technological knowledge to raise the scientific and cultural level of all citizens."

Finding Postdocs and Grad Students

Good postdoctoral researchers (and, for academic researchers, good graduate students) can make a big difference in a lab. They bring new ideas and expertise into a research community and contribute to a lab's productivity and prestige. But those students won't apply to be part of a program if they don't know it exists. I've worked with researchers who have told me how grad students applied to work in their labs after first reading mainstream media accounts of a lab's work.

Getting Interest from the Private Sector, Policy Makers, and Nongovernmental Organizations

Not every research project will be of interest to the business, government, and nonprofit sectors, and not every researcher wants to work with them. But they can be valuable. For example, if new research findings have ramifications for bridge builders, landfill operators, or policy-making bodies that oversee those fields, it makes sense to get that information into the hands of people who can use it as quickly as possible. The findings may not be ready for immediate application; they may simply offer a glimmer of possibility that needs to be explored further. But sharing that information with the broader world creates opportunities for research partnerships (and potential funding) that may not otherwise come to light.

EXPLAIN THE PROCESS

Once you've laid out the potential benefits of researchers promoting their research findings, it's important to explain precisely how you can work with them moving forward. In other words, you need to tell the researchers what you need from them—and why you need it. Whenever I meet with new researchers, I give them "the spiel," which is my name for the following five pieces of advice.

Be Willing to Talk to Reporters

This is a threshold issue, meaning that if researchers can't do this, it will be impossible for a PIO to work with them. Researchers who promote their work need to know that, at some point, they may have to actually talk to a reporter. A good PIO can get reporters interested in a research story, but the reporters will want to speak to the scientists who actually did the work. If researchers aren't willing to talk to reporters, there's no point in promoting their work. That said, let the researchers know that you're willing to provide them with media training to make them more comfortable and confident when discussing their work with non-experts. (We'll explore media training in detail in chapter 5.)

Let Me Know What's Going On

This is also a threshold issue. A PIO can't promote research that he or she doesn't know about. Ask researchers to keep you abreast of what they're working on. This task focuses on two characteristics of the researcher's work: timeliness and content.

Timeliness is important because the research community and the news community have very different senses of time. If a journal article came out six weeks ago, researchers think it is still brand-new. After all, it hasn't really had a chance to penetrate the intellectual marketplace yet. However, six-week-old news is ancient history to most reporters. Some science-beat writers may have a more generous criterion for timeliness, but many science news stories these days are written by general-assignment reporters, and they like their news to be new. In practical terms, this means that researchers should give you a heads-up about journal articles as soon as the article has been accepted. Most journals have moved to an online model, which means that a paper often appears online within days or weeks of being accepted. You want to be prepared to promote that research as soon as it becomes publicly available online (or even earlier if there's an embargo, which is discussed in chapter 3).

The other common window of opportunity for promoting research findings is when work is presented at a peer-reviewed conference. The importance of conference papers varies considerably depending on the conference and the research discipline: for instance, they are considered extremely important in computer science but unimportant in biological engineering. Talk to researchers to find out how important conference papers are in their field. If they are important, ask researchers to tell you about forthcoming papers at least two to three weeks before the conference. This advance notice allows you to promote the paper at least a week before the conference starts, which has two benefits. First, the researchers are still in their labs or offices instead of at a distant conference location. This makes it easier for reporters to reach them. Second, if you are successful at drawing attention to the research, you will have created additional opportunities for the researchers to network while at the conference. Recent media coverage of their research increases the likelihood that people will show up for their presentation or will approach them by the coffee table with questions about the work. And networking is what conferences are all about.

A Note on Controversial Subjects

I'm in favor of promoting good work, even if it's about a subject that some find controversial, such as evolutionary biology or climate change. I think that making decisions based on fear (e.g., not promoting climate-change research because some people may dislike the research findings) ultimately leaves you paralyzed. If the science is good, then it is worth promoting. However, if there is a possibility that the research may ruffle feathers, it's always best to notify your higher-ups. They may offer insights into problems or angles you'd overlooked, and they will always appreciate being informed about communication efforts before receiving any potentially angry e-mails.

In addition to timeliness, you want researchers to think about content before they contact you with news about their research. To frame it in a slightly different way, you probably do not want researchers to tell you about every single paper they put out or conference proceeding they attend over the course of a year. Researchers think that every paper they publish has value; otherwise they wouldn't have put their names on it. But some papers are more important than others. I ask researchers to bear four things in mind when determining whether to let me know about a forthcoming paper:

1. Will this paper make other people in your field sit up and take notice? If so, please let me know.
2. Does this paper have practical applications? For example, will it help people be healthier, help them save money, or make a process more energy-efficient or cost-effective? If so, please let me know.
3. Is the paper really cool? If you told your neighbor about it, would they say "Wow!" If so, please let me know—I'll probably say "Wow!" too.
4. If you're not sure about any of those things, but you think one of them is a possibility, please let me know.

Help Me Get It Right

Once researchers have told you about a forthcoming paper—and have hopefully sent you a proof copy—you'll need to meet with them to learn more about it. It's important to try to read the paper, but odds are excellent that you won't fully understand all of it. At my institution, I've worked with researchers on findings ranging from forensic entomology to computer malware. There's no way I can familiarize myself with jargon from so many different fields or develop the necessary expertise to place all of these findings in context. So I ask a lot of questions.

When writing about research findings, I usually start by finding out what question or challenge researchers were setting out to address. If scientists phrase things in technical language, I'll have them define the terms. Then I ask them why they found this problem interesting. Sometimes the answer is pure intellectual curiosity, but usually the research question is one element of a much broader scientific question. Science is an iterative process, and the findings from a single research project may move us incrementally closer to understanding the genetic basis for a disease or the factors affecting the efficacy of antibiotics, for example. If I can get researchers to place their work in context, it becomes much easier to explain the relevance of their work to a lay audience.

Once I think I have a handle on their work, it's my job to write a news release or blog post that explains the research in terms that are accessible to a non-expert audience and that highlight what is interesting or important about the work. Then, before anyone else sees it, I send the piece to the researchers for review. I make sure to stress this part of the process to researchers, because I want them to know that getting the facts right is as important to me as it is to them.

Help Me Find the Right Outlets

Many researchers think that PIOs are primarily interested in placing stories in mainstream media such as *USA Today* or the *New York Times*. You need to tell them there is more to it than that. To encourage researchers to consider promoting their findings, I tell them I'm also interested in pitching stories to news outlets that are read primarily or exclusively by their research community. There are magazines, websites, blogs, and newsletters that focus on almost

A Note on Inter-Institutional Research News

If someone wants you to promote a research paper that has authors from multiple institutions, make sure you coordinate promotion with your counterparts at those institutions. The first author's institution often takes the lead in promoting the research, but that is not always the case. You'll want to determine who is responsible for writing a news release, who will pitch the story to reporters, and so on. This avoids any duplication of effort and keeps you and your counterparts from annoying reporters with multiple pitches on the same story.

every conceivable field of study; you just have to find out what they are. Ask the researchers where they go, outside of the peer-reviewed literature, to keep track of developments in their field. Researchers often find the prospect of appearing in these outlets significantly more interesting than the idea of seeing their work written about in mainstream media.

This Will (Probably) Be Painless

It's important for PIOs to tell researchers that promoting their own work doesn't have to be particularly time-consuming. Part of a PIO's job is to make the process as effortless as possible for the researchers he or she is working with. PIOs are there to help researchers promote their work, not to distract them from it.

RESEARCH-RELATED NEWS (THAT ISN'T RESEARCH)

So far, we've discussed how to find good stories that are about research. But if you work as a science PIO, you'll find yourself faced with a lot of stories that are research-related but don't include any research at all. Examples include new grants, research awards, or the hiring of new researchers. These are stories that are often presented to a PIO as a "great idea" or a "big deal" but are often of limited interest to an external audience. This can put you in a tricky situation because you don't want to offend anyone. And sometimes a story with limited news value is still worthwhile—if it can help you reach a target audience and advance your institutional goals. A story that might only get coverage in the local newspaper's business section could still be worth doing if your institution is trying to forge partnerships with the local business community. Let's look at these research-related subjects individually.

Grants

I'm mentioning grants first because they are essential to the productivity of almost all research institutions, and because I approach grants much the same way that I approach other research-related subjects: warily. It's hard to get external audiences interested in a grant because the researchers who received the grant *haven't actually done anything yet*. Instead, the researchers have identified a problem or challenge and have made a convincing argument that they might be able to shed some light on it. The argument was apparently so convincing that someone was willing to give them money to do the relevant work.

So what?

Very few reporters are interested in writing about grants. Instead, reporters are interested in the research findings that stem from the work that grants support. Those findings usually don't appear until years after the grant is awarded. Research grants are important, *but being important is not the same as being interesting.* For example, funding for cancer research is enormously important. But lots of researchers at lots of institutions get lots of grants to study various aspects of cancer. No one is going to write about every single one of those grants, because the grants aren't interesting in themselves. The grant will get written about when and if it leads to interesting findings.

However, promoting grants can yield useful results. It took me a while to figure this out. When I started working as a PIO, I thought writing about grants was a waste of time. But if the grant was big enough, I had to write about it anyway. I eventually learned that promoting grants can provide very practical benefits for the researchers involved. I had written about a reasonably significant research grant, and while most reporters had ignored it, a few specialized outlets did pick up the story. It was through one of those news outlets that another researcher (I'll call him "Dr. Z") read about the work that "my" researcher ("Dr. A") would be doing with his new grant. Dr. Z was working on a similar—but different—research project on the other side of the country. Dr. Z contacted Dr. A and offered to share his large data set. Dr. A agreed immediately and said he would return the favor once he'd compiled his own data set. Dr. A was off to a running start on his research project, and both researchers would ultimately get larger data sets than they anticipated. The real winner, of course, was the research itself; a more robust data set is a good thing.

Dr. A called to tell me his exciting news. He said he was sure he wouldn't have gotten the data set if we hadn't decided to promote his grant, for the simple reason that Dr. Z's research was just different enough from Dr. A's that they would probably not have met under other circumstances—and certainly not in such a timely way for Dr. A's work. It dawned on me that *this* was a great reason to promote grants: to raise awareness in the research community of emerging research initiatives, thus creating opportunities for formal or informal collaboration. The lesson here is that promoting grants can yield positive results for both your institution and the researchers you're working with. However, you need to recognize the limitations you're up against.

Decide whether to promote a grant award based on the research that the grant will support. If the work is interesting, you stand a better chance of getting reporters interested. Also, focus on news outlets that are more likely to write about a grant award. These are often local news outlets and ones that focus on specific research areas. (I'll say more about finding the right media outlets in chapter 3.) Lastly, make sure to manage expectations. Congratulate the researchers on their grant, but make it clear that most news outlets aren't going to write a story about it. It's always better to under-promise and over-deliver, if possible.

Inventions and Applications

Inventions and research applications can be difficult to promote because they often present thorny legal and business questions that most PIOs will need help navigating. Research institutions often face limits on the extent to which they can endorse products. Make sure you know what those limitations are before agreeing to help a researcher promote an invention. For instance, I work at a public university that is not allowed to endorse products or services in any way. If a researcher at the university built a better mousetrap, I would have to walk a fine line: describing the importance of the researcher's work without saying that the new mousetrap is a great product. In situations like this, it makes sense to reach out to your institution's attorneys. They have legal expertise that can help you determine what you can and cannot say—or whether you shouldn't say anything at all.

Another challenge inherent in promoting inventions and new applications is the need to protect the researcher's (and the institution's) intellectual property. In the United States, inventors have a twelve-month window of opportunity for filing a patent application. The clock starts ticking as soon as an inventor's work is discussed publicly—but only if the public statements are "enabling," meaning that they are sufficiently detailed to allow someone else to reproduce the invention. (Other things can also start the patent clock, but this is the only one that's relevant to PIOs.) This means you need to make sure that the researchers have taken all the steps necessary to protect their intellectual property before you agree to promote the work. Both the timing and the wording can be tricky when dealing with intellectual property, so I'd recommend developing a relationship with your institution's tech transfer office or with the personnel in the legal affairs office who handle intellectual property.

Finally, some invention and application stories are simply unlikely to garner a great deal of attention. For example, an exciting new product may have ties to work done years ago at your institution, but reporters are unlikely to focus a story on the years-old research that led to the new application. If you think the story is strong enough, you can still try to promote it, but you will want to limit expectations for the impact it will have. Conversely, recent research may lead to a new product that is so specialized or technical that very few people will understand what the product does, much less why it's important. If that's the case, you may want to focus exclusively on pitching the story to news outlets that focus on the researcher's specific field of study.

Partnerships

Much like grants, new partnerships between organizations are often of interest primarily to the parties involved—and no one else. When determining whether and how to promote a new partnership, focus on how the partnership will benefit your institution. If you can clearly outline those benefits (e.g., your institution is receiving significant funding or new laboratory equipment), it may be worth promoting. If the partnership only has *potential* benefits, or the benefits are poorly defined (e.g., it will "foster collaboration"), you may not want to promote it. A good reporter will tear those vague statements apart.

The news organizations most likely to cover partnership announcements are local media outlets, particularly if both partners are located in the same area. Otherwise, let the nature of the new partners drive your media outreach strategy. For example, business news outlets are more likely to cover partnership announcements involving private businesses, and higher education news outlets are more likely to cover partnerships between universities. The endorsement concerns that apply to promoting inventions also apply to promoting partnerships with private businesses. You want to highlight how your institution's partnership with a business is good for the institution, without appearing to endorse that business's products or services.

Awards

Award announcements are usually a losing proposition. They are extremely prestigious within a researcher's field but are rarely of interest to anyone outside of that field. Generally, it is best to pitch the news directly to any news outlets that focus specifically on the relevant field of study. If you pitch it to mainstream news outlets, you're basically just spamming them. There are, of course, exceptions. If a researcher wins a Nobel Prize or is named to the National Academies, that is certainly worth promoting. Awards and honors below that level usually won't arouse much mainstream media interest.

New Hires

Unless the person you're hiring has an incredibly high profile—on the level of a Nobel laureate—almost no one is going to care about your institution's HR decisions. If the new hire is essentially a celebrity in his or her field of study, notify any news outlets that focus on that field. Otherwise, announcing a new hire is probably a waste of your time (and your audience's time).

Special Events

Certain special events can be of significant interest, particularly to local news outlets, but that depends on what sort of event it is. Reporters will rarely turn up for a symposium or a technical conference, but they may come out if significant new research findings will be announced or if there is an especially high-profile speaker. When asked to promote an event of this kind, get as many details as possible.

If new research findings will be unveiled, be sure to contact reporters who cover the relevant beat well in advance so they can make travel arrangements, if necessary. If there is a high-profile speaker, give local reporters advance notice (out-of-town reporters are unlikely to attend, unless the speaker is making an announcement of some kind), and remind local TV news crews the day before the event, because they usually don't commit to covering an event more than a day or two in advance.

The other type of event is the official groundbreaking for a new facility, such as when an institution is preparing to begin construction on a new research lab. These events are likely to be covered solely by local news outlets. Before pitching a story to reporters, make sure you're able to explain why the new facility is important. For example, if the facility will house new equipment that will boost genetics research on the origin of specific types of cancer, you'll want to tell reporters that. This helps them place the event in context so that instead of writing about a new construction project, they can write about a new center for cancer research (which is significantly more interesting).

A science PIO's job doesn't end with inviting media to an event. A PIO will need to make sure there is room for reporters, photographers, and camera operators at the event (with a good line of sight for the photographers and camera operators). You'll also want to make sure the venue will have a "mult box"—a device that allows reporters and camera operators to plug directly into the event's audio system so they can get good audio recordings of whoever is speaking at the event. You'll also want to make sure that representatives of your institution are available to talk with reporters (on camera, for TV reporters) either before or after the event. (And you'll want to make sure your institution's representatives are prepared to talk to reporters—see chapter 5.)

FINAL NOTES

Everything I said above is true. PIOs should look for research that is important, practical, or fun. But you should also seek out research that gives you an opportunity to promote your institution's goals. At the very least, you should write about the research in a way that is consistent with your institution's goals and that will help you reach the right audiences. (More on reaching your audience in the next chapter.) Also, you'll note that I don't mention press conferences in this chapter. That's because I think they should be used sparingly. (I discuss press conferences at length in chapter 3.)

KEY POINTS

- Know your institution's goals, and keep those goals in mind when deciding what to write about.
- Talk to management, and familiarize yourself with the organization from top to bottom.
- Build relationships with researchers; they're the ones who have stories to tell.
- Be able to explain "selfish reasons" for researchers to promote their findings, and bear those reasons in mind when choosing, writing, and promoting stories.
- Let researchers know what you need from them: they should keep you in the loop on their work, help you get it right, and advise you on where the stories should run.
- Remind researchers that you're there to help them promote their work—not to distract them from it.
- Be discriminating when it comes to promoting research-related news (other than research findings): be aware of potential legal issues; differentiate between "important" and "interesting" (i.e., newsworthy); and remember that small stories can still be valuable if they'll help you reach a key audience.

2

WRITING STORIES

After you've found a good research story, you need to translate it into language that people will want to read. This chapter will address questions every PIO should ask when telling a story, whether in the form of a news release, a blog post, or a feature article. Who is your audience? How long should your story be? How can you craft an opening paragraph that will grab people's attention? And what should you do to make sure that you got your facts right?

BEFORE YOU START WRITING

Before you start writing, you need to know what you want to convey with the piece. This is particularly important for PIOs, since they are often tasked with trying to accomplish multiple things at the same time—most often determined by what their "client" wants. For a science PIO (like me), there are usually two clients: the researcher whose work you are promoting, and the institution you work for (in my case, a university). I need to define the goals of the researcher and the university before I begin writing because the goals tell me who the target audience should be, and the audience shapes the way I write.

Identify Your Key Audiences

Research institutions and researchers usually have different, but overlapping, goals when it comes to science communication. I'll use universities (rather than, say, federal labs) as an example, since I work at one. Universities want to raise their profile in a positive way with a wide variety of external audiences: potential students, potential faculty, alumni, federal funding agencies, state legislators (if they're public universities), business partners, and businesses they want to partner with in the future. Note that most of these audiences are outside the sphere of the science community.

Researchers, on the other hand, want to attract promising grad students, have their papers cited, identify funding opportunities, curry favor with federal funding agencies, and create opportunities for formal or informal collaboration with other researchers. Universities also share these goals because anything that reflects well on a university's faculty also reflects well on the university itself. But please note that the audiences you need to reach to accomplish any of the researcher goals are primarily within the science community.

Because I'm a university PIO, I try to accomplish the goals of both my university and the researchers I work with. That means simultaneously writing for an audience of teenagers (i.e., potential students) and an audience of PhDs (i.e., potential collaborators). I do that by trying to write clearly, not talking down to my audience(s), and highlighting whatever is interesting or important about the research. Specifically, I write for an imaginary person whom I think of as the "intelligent non-expert." The intelligent non-expert is familiar with basic scientific concepts but not with the intricacies of whatever subject I happen to be writing about. This means I keep jargon to a minimum and explain any that I do use. Jargon is not inherently bad—it allows experts to convey large amounts of information in concise terms—but it can scare off readers, particularly if you use it early in a story. I'll discuss jargon at greater length later in this chapter.

What I *do* want to include early in the story is an explanation about why the research matters. If the work moves us closer to more efficient lasers, targeted drug delivery, or understanding the genetic basis of a disease, you should say so. Don't oversell the research, but place it in a context that people can understand. If it is impossible to explain the work's relevance to a practical application, then at least put the findings into context with previous work. For example, if the research resolves a scientific question that was first asked twenty years ago, say that. And since science is an iterative process, it may be worth discussing what new questions the research raises. Questions can be as interesting as answers, if they're really good questions.

If you've written clearly and explained the research's relevance, then the piece will likely be interesting and understandable to at least some (if not most) of the audiences you want to reach. Experts in the field may be annoyed that you didn't include all the nitty-gritty details, but you can point them to the paper; they're the audience the journal article was written for in the first place. (Note: Always remember to include a link to the paper in your story. If the paper is not open access, meaning reporters won't be able to access it, make sure you are able to provide reporters with a PDF of the paper.)

News Releases versus Blog Posts

Another thing you need to decide before you start writing is whether you're writing a news release or a blog post. News releases tend to follow a certain formula, whereas blog posts offer writers more freedom in terms of narrative style and length. If your institution doesn't have a blog, the decision is easy—clearly you'll be writing a news release. But if your institution does have a blog, the decision isn't that much more difficult. If your primary goal is to get reporters to cover something, you should write a news release. If you're trying to reach public audiences directly, write a blog post. Also, make sure to familiarize yourself with your organization's review process for news releases and blog posts. If there are multiple layers of review, such as those found in many federal agencies, the review process can be fairly time-consuming. If you want to issue a release or blog post in a timely way, you'll need to budget in the time necessary for the internal review.

It's also worth noting that, in a sense, the construct of "news release versus blog post" presents a false choice, since there are other options available, such as feature stories for your institution's website. Features tend to allow the same narrative flexibility as blog posts. This section offers general guidelines, which you can adapt according to your needs.

LEADS AND LEDES

A news release is not the same thing as a news story written by a reporter. For example, institutions rarely want PIOs to interview multiple third-party sources to provide outside perspective in their news releases. A news release is a summary that reporters can use to determine whether they want to write a news story. However, a news release needs to have most of the same elements as a news story. In part, this is because some news outlets will run a news release as if it were a news story. But more importantly, this is because you want people (especially reporters) to actually read the news release. One

The News Release Is Not Dead

Some people say the news release is "dead." Their argument is that science PIOs don't need mainstream news outlets anymore and should focus their time and energy on writing blog posts and using social media to communicate directly with the public.

I disagree.

Science communication is not a zero-sum game. PIOs can reach out to news media via news releases while also reaching out to audiences directly via blogs and social media. More importantly, mainstream news outlets can reach more people than you can. Blogs and social media are great (more on them in chapter 6), but the people who read your blog or follow you on Twitter are people who are already interested in your institution. If you want to reach people who aren't already interested, those tools may not be very effective.

For example, if you're writing about astrophysics research, one of your goals will be to reach astrophysicists, and they probably don't read your institution's blog (unless you work at NASA). However, if you can get a reporter at *Scientific American* or *Astronomy* to write about the research, the astrophysics community will definitely take notice. And I find news releases to be an invaluable tool when I'm pitching to reporters (which is what chapter 3 is all about).

Finally, news releases can also help you disseminate news about your institution because many news outlets will pick them up and run them verbatim (or almost verbatim) on their websites. These outlets may be large news aggregators, such as phys.org, or local newspapers with limited news staff.

of the features common to both good news releases and good news stories is a good lead.

You need to find a lead (rhymes with greed) that clearly establishes a news hook, telling readers what's interesting about the story and why you're telling them about it now. Whether you're writing a news release or a blog post, there is nothing more important than giving people the lead right away—because *you need to give the reader a reason to keep reading*. This is easier said than done, particularly when writing about complex scientific subjects. Ideally, your lead (the news hook) will be in your *lede*, which is the term used to describe the first paragraph of a news release, blog post, or news article.

There are many ways to begin a story. Finding the right opening line is important if you want your audience to keep reading. It often takes me as long to write the lede as it does to compose the whole rest of the story. As Tim Radford, a former editor at the *Guardian* newspaper, wrote in his famous "Manifesto for the Simple Scribe," "There is always an ideal first sentence." But how do you find that ideal first sentence?

Ledes for News Releases on Research Findings

Writing the lede for a news release is pretty straightforward. It should basically be a classic news story lede. That means you want to keep it short—no more than two sentences. And you have to provide readers with a summary of what the news release is about and why they should care. Basically, you want people to be able to determine, at a glance, whether they should keep reading.

Although writing a news release lede is straightforward, it is not always easy. Many scientific findings come with a host of qualifiers. For example, you might not be able to say that this is the first time researchers have been able to produce gold nanoparticles. Instead, you might need to say that this is the first time researchers have been able to make a *specific kind* of gold nanoparticle, under *these specific conditions* using *this specific technique* with *this specific technology*. That can bog down anyone's opening sentence, and you still haven't told readers why they should care.

Here's an example of a lede I wrote for a news release on research findings:

> Researchers from North Carolina State University have designed software that allows them to map unknown environments—such as collapsed buildings—based on the movement of a swarm of insect cyborgs, or "biobots."

This lede is short but still manages to explain, in broad terms, what the researchers did ("designed software") and why they did it (to enable them to map unfamiliar terrain). It also hints at a specific reason that readers should care—namely, to help people determine whether a collapsed building is safe. In addition, it highlights an interesting aspect of the research (insect cyborgs) and clearly states where the work was done (North Carolina State University).

This is a good lede because it's eye-catching and provides sufficient detail to help reporters determine if they want to read more. Here's another example:

> Researchers from North Carolina State University have developed a new technique to identify the proteins secreted by a cell. The new approach should help researchers collect precise data on cell biology, which is critical in fields ranging from zoology to cancer research.

This lede is less exciting, but I still think it's a good example. The work being described was incredibly complex and loaded with scientific jargon that many readers wouldn't understand. This lede manages to give readers a general idea of what the researchers did and why it might be important.

To help you understand why it was a challenge to translate research findings into accessible language for that lede, I'll tell you the title of the related journal article: "Targeted Proteomics of the Secretory Pathway Reveals the Secretome of Mouse Embryonic Fibroblasts and Human Embryonic Stem Cells." This sort of technical language would likely have prevented most reporters from reading further, because it wouldn't be clear to them what the research was or why they should care. However, someone else wrote an even better lede about the same paper:

> Chemical communication between cells keeps tissues functioning and systems coordinated, but eavesdropping on the conversation is challenging. Now, researchers have developed a technique to identify signaling proteins before they leave the cell. The method could help determine which cells are sending which messages—a useful tool for analyzing the interactions occurring in the mixed populations in tissues. One possible application could reveal the cues that control stem cells—an insight that researchers hope could be applied to healing damaged tissues.

Freelance science writer Marissa Fessenden wrote that lede for an article in *Scientific American*—and I really wish I'd written it myself.

She found a way to provide more detail about the work while still using accessible language. I prefer to write shorter ledes for news releases, so I would have pushed the last two sentences of that lede into the second paragraph. But I wouldn't change a word of what she wrote.

At the other end of the scale, here's an example of a poorly written lede:

> In an assessment of the nasal floor configurations of the available and sufficiently intact, if still incomplete, paleoanthropologists from Institute of Vertebrate Paleontology and Paleoanthropology (IVPP), Chinese Academy of Sciences, Washington University and University of Missouri, found that archaic *Homo maxillae* from eastern Eurasia seem to have a prevalence of the bi-level pattern similar to that seen in the western Eurasian Neandertals, while early modern humans from eastern Eurasia mostly exhibit the level floor pattern predominant among early and recent modern human populations, indicating that bi-level nasal floors were common among Pleistocene archaic humans, and a high frequency of them is not distinctive of the Neandertals as thought before.

I first saw this lede at a conference presentation by *Wired* reporter Nadia Drake on how to write useful news releases. This was given as an example of what not to do. The entire lede consists of one immensely long sentence, and the language is not accessible to nonexperts. As a result, reporters have no idea what the researchers discovered or why it's interesting. Ledes like this one highlight the need for good science PIOs.

Ledes for News Releases about Grants or Other Research Activities

Writing a lede for grant or event announcements follows the same principles as those for a release on research findings: explain what the release is about and why reporters should care. But these ledes are a little more challenging because it's often harder to get reporters interested in grant awards or event announcements. When I write about grants, I try to talk as little about the grant as possible. You should dispense with the grant in a single paragraph, saying

Be Honest

The single most important thing to keep in mind when writing is to be honest. If you want reporters to trust you and take you seriously, you should never overhype research findings. It's fine to present the work in the best light possible, as long as you're telling the truth. But if you mislead your readers, you will lose their trust. Once you've lost your credibility, reporters won't want to work with you—and that will make it virtually impossible for you to do your job.

how much the grant is for, its duration, and where the funding came from. Here's an example:

> North Carolina State University is taking the lead on a five-year, $7.3 million "citizen science" initiative funded by the National Science Foundation. The goal of the program is to give science teachers and students the opportunity to engage in meaningful scientific research while improving the educational success of both teachers and students.

This lede gives readers all of the basic information about the grant, as well as a concise overview of what the researchers are trying to accomplish.

News releases about grants should also highlight the problem or challenge that the researcher will use the grant funding to address. This doesn't have to happen in the lede, but focusing on the problem can make a grant announcement a little more interesting:

> A new grant to North Carolina State University and several partners could make installing rooftop solar energy systems much less expensive and time-consuming.

My North Carolina State University colleague Nate DeGraff wrote this lede, and it's a good example of keeping it short and focusing on the challenge that the researchers are hoping to address. Once the lede grabs your attention, you can give people additional information about the grant itself. In this instance, DeGraff covered all of the grant details in a single sentence in the second paragraph.

These same concepts apply to news releases about events, institutional partnerships, or anything else a research institution may want to publicize. Here's an example of a release to promote an event:

> Law enforcement officers from around the United States are coming to North Carolina State University to participate in a week-long crime scene investigation workshop, which involves solving a series of mock murders while learning about state-of-the-art technology in the field of forensic science.

An NC State PIO named Caroline Barnhill wrote this lede. She did a great job of highlighting precisely why this event might be of interest to reporters, eschewing technical language while accurately describing the goals of the event.

LEDES FOR BLOG POSTS

Blogs give writers an enormous amount of flexibility. While news releases tend to be somewhat formulaic (more on that anon), blog posts can be written any number of ways. They can be funny or sad. They can be written in the first person. They can be human interest stories. They can be incredibly short, or lengthy, in-depth features. But they all need a good opening line.

You can always write a blog post in the traditional news style, like a news release, but it can be more effective (and more fun) to explore other ways of opening your story. Here are four different approaches that science writers use to draw readers into a story. This is not an exhaustive list; some of the categories can overlap, and there are many others that I don't touch on at all. But hopefully this will give you some ideas that you can use to explore different narrative styles.

Mystery

Some science stories begin with a line that simply begs to be explained. I don't mean that they are written in technical language that can't be understood; I mean that they capture a reader's imagination and almost force the reader to find out what is going on. Science writer Carl Zimmer is particularly adept at this. Here's a great example: "In 1941, a rose killed a policeman." That's the opening line of a blog post that Zimmer wrote on National Geographic's *Phenomena* site. After I read that first line, I had to keep reading. Ultimately, the piece was about antibiotics and the microbiome. Some readers might have been put off if the story had begun with a reference to the microbiome, but by the time you got to the nitty-gritty details, the story had built up enough narrative momentum to propel readers through to the finish line.

Humanize It

It might be difficult to get people to read a story about microbiology. It could even be a challenge to get people to read about something as eye-catching as a crime, if the crime happened some place far away. But if you open the story by making it about a person, you can usually get someone's attention. Pulitzer Prize–winning reporter Deborah Blum used this approach powerfully in a piece she wrote for her Wired.com blog *Elemental*. Here's the lede: "Your daughter died. Your daughter died thousands of miles from home." Any reader, or at least any reader who is a parent, can immediately connect to the story. Even if they don't yet know exactly what the story is about, they recognize it as every parent's nightmare. That connection impels readers to keep reading.

For a less direct example, I'll point to a piece that Melinda Wenner Moyer wrote for *Scientific American*. Here's the lede: "David Gems's life was turned upside down in 2006 by a group of worms that kept on living when they were supposed to die." Gems, Wenner Moyer quickly explains, is a researcher who uses a specific species of roundworm to study the biology of aging. But by making the first sentence focus on Gems, rather than his work on cellular damage caused by oxidation, Wenner Moyer pulls the reader in. Many readers won't be initially interested in (or perhaps even understand) oxidation, but they can identify with a person. This lede could also be used as an example of a "mysterious" way to grab a reader's attention.

Make It about the Reader

People love reading about themselves, so making the story be about the reader is often a good bet—especially if you're writing about something that touches on their health or can save them time or money. This technique may be overused (and is often misused), but it's not necessarily a bad idea. Gisela Telis wrote a piece for *ScienceNow* that begins like this: "If you were plagued by pimples in your teen years, you may have bacteria to blame—but not all of them." As her story notes, more than 80 percent of people in the United States have experienced acne and can therefore relate to her lede. Former acne sufferers may want to read it to find out what was behind their condition, and those still dealing with acne will keep reading to see if there's anything they can do about it.

External News Hook

If you can tie your story to something that is the subject of significant public attention, odds are good that you can tap into that interest—even if the subject is something silly. I wrote a blog post that begins this way: "In early February, astrophysicist Neil deGrasse Tyson said on Twitter that the superhero Thor's hammer (aka Mjolnir) 'weighs as much as a herd of 300 billion elephants.' News outlets pounced on this, and the news was quickly circulating online. Sadly, Tyson was wrong." I did not do this to pick on Tyson, whom I admire. I wrote that lede because I wanted to talk about physics and materials science, and tapping into a recent news story about a high-profile scientist (as well as the enthusiasm of comic book superfans) was a great way to do it. It worked.

WRITING THE BODY OF A NEWS RELEASE OR BLOG POST

Coming up with a good lede is often the hardest part of writing a news release or a blog post, because it has to be concise while being catchy enough to grab a reader's attention. Writing the rest of the story—the so-called "body" of the piece—is usually much easier, but there are still certain things you want to make sure you include.

The Body of a News Release

The body of a news release about research findings should summarize the work, but it should not be an exhaustively detailed description. Journal articles can be dozens of pages long, and there is no way to include all of the relevant information in a release. However, the body should convey some basic facts about the work, including (usually in this order):

- an overview of the question or challenge that researchers were setting out to address
- a description of the findings
- a statement of why the findings are important (fleshing out what you wrote in the lede)
- the methods used in the study
- the limitations of the study (be honest!)
- future directions for the research
- the names and affiliations of the researchers
- and, if applicable, who provided funding for the research

You also should include at least one quote from one of the researchers. This adds color to the release and can offer some insight into the researcher's perspective. Also, a good quote can signal to reporters that the researcher is worth interviewing.

A former editor of mine told me that a good quote should say something that you can't say yourself. In other words, the quote should not be a straightforward statement of fact. It should offer perspective or highlight some way that the research could be important. For example, a PIO shouldn't write a sentence saying: "Dr. Smith thinks that this work will pave the way for new research into the cellular processes associated with Disease X." Did the PIO read Dr. Smith's mind? No. So the only way the PIO would know what Dr. Smith was thinking would be if Dr. Smith said so. In that case, a good quote from Dr. Smith would allow the PIO to write this sentence instead: "'I think this work will pave the way for new research into the cellular processes associated with Disease X,' says Dr. Jane Smith, an associate professor of biology at NC State and senior author of a paper describing the work." (On first reference, it is usually best to list a researcher's full name, title, and association with the research. However, if this disrupts the narrative flow or makes for an awkward paragraph you can include the title and related information lower down in the piece.)

It is also helpful to include any relevant grant numbers (e.g., NSF grant DGE-0750733) if research was supported with funding from a federal agency. Grant numbers can disrupt the flow of text, so I recommend placing this information near the bottom of the release. Federal funding agencies encourage research institutions to do this because it makes it easier for the agencies to track communication efforts related to their grants.

If the research findings are being published in a peer-reviewed journal, the related news release should also include a hyperlink to the online version of the article and the article's DOI. The DOI, or digital object identifier, is a number associated with a specific journal article and can be used to search for the article online. A DOI looks something like this: DOI:10.1063/1.4821040.

If the research findings are being presented at a conference, there will be no DOI and the paper may not be online. However, it's worth checking with the lead researcher to see if they are able to post the paper online. If so, make sure to include a hyperlink to the paper in the body of the release.

For news releases about grants, focus on who the researchers are and the question or problem they will be investigating under the grant. You should also check to see whether the researcher is taking a particularly interesting approach to solving the problem. If so, include that information in the release. It's also a good idea to explain how the researcher's approach fits into the existing body of knowledge about the subject.

News releases about events, awards, or partnerships are essentially announcements. Keep them short. Leave out extraneous details and focus on the who, what, when, where, and why. For events, make sure you say whether the event is open to the public, if it's free, and whether people will need to get tickets in advance, as well as providing information about parking. If the event is not open to the public, make sure that members of the news media can attend. Otherwise, you shouldn't do a news release for it. Note: A sample research news release is included at the end of this book (appendix B), as is a sample grant announcement (appendix C).

The Body of a Blog Post

A science blog post needs to achieve the same goals as a science news release: explaining the work and telling why it is interesting or important. However, a blog post gives you far more flexibility in how you tell the story. The body of a story unfolds from the lede, and I've already outlined several different ways to open a blog post. For example, a blog post can be written like a feature article, focusing on the people who are doing the research rather than the research itself. Similarly, a blog post can examine the back story of a research project, rather than focusing solely on the project's findings. In 2013 I wrote a blog post about NC State researcher Peter Ferket, who is an expert on poultry nutrition. He was a co-author on a paper published in *Science* that studied genetic drivers of obesity, and I was wondering how a chicken expert had ended up co-authoring a genetics paper led by a group of Harvard researchers.

I interviewed Ferket and learned that the Harvard-led team was looking for a way to determine how much energy their lab mice were absorbing from their food. Ferket, who had worked on similar problems in poultry, devised a new technique for assessing the metabolism of mice that involved blowing up mouse feces and seeing how much energy the explosion produced. I decided to write about the research challenges that led to the creation of Ferket's new technique, rather than focusing solely on the findings discussed in the *Science* paper. This made for a pretty interesting blog post and highlighted Ferket's role in the research project far more clearly than would have been possible if I'd written a conventional news release. It also caught the eye of a science reporter who wrote it up for the science news outlet *Nautilus*.

In short, blogs give you the freedom to explore different narrative styles. You can write 2,000-word features or 200-word blurbs. Your ultimate goal is to get people interested in the work being done at your institution. Keep that goal in mind, but be willing to experiment with different storytelling techniques. Look at news articles or blog posts that you admire from other news outlets. Pay attention to how those writers are telling their stories. Then try to mimic those storytelling styles in your own writing. When you're done, step away from the desk and take a break. When you come back, take a long look at what you wrote. If you work with other writers, have them look at what you wrote. If it's good, go with it. If it needs work, either try again or throw it out. But don't be afraid to try something new.

HOW LONG SHOULD THIS BE?

A news release needs to be long enough to explain the research but short enough to hold a reporter's attention. The main goal for a news release is to interest reporters so they'll want to follow up and write their own articles on the research. Reporters get approximately one

Do I Need to Read the Paper?

Should science PIOs read the journal articles they are writing about? Yes. But don't panic if you don't understand everything you're reading. Some PIOs have training as scientists, but many don't. And even if you're trained as a scientist, that won't necessarily help you. Having a PhD in astrophysics probably won't be much use if you're reading a journal article about plant biology.

I am not a scientist. And although I read journal articles, I often don't understand what I'm reading. It may be clear prose to the experts, but it's a shibboleth to me (and if you don't know what a shibboleth is, you know how I feel when I read phrases like "rectification using multiheterojunction"). Yet I've made a career out of writing about science and have never had to run a correction related to the scientific content I've written about. This is because I ask researchers a lot of questions: "What questions were you trying to answer with this research? Why? What was your methodology? What were the key findings? What new questions did this research this raise?" And every time I don't understand the answer, I ask them to explain it. The researchers act as translators, walking me through the paper step-by-step. This dialogue is essential for any non-expert (like me) who wants to write about a paper. It allows me to understand the content and context of the research in the paper.

That said, PIOs should try to read the papers they're writing about. Once in a while, they're actually written in prose that is accessible to the lay reader, and you can usually glean useful information from the abstract, introduction, and discussion sections.

zillion news releases and pitches per day. If you send them a 1,000-word summary of research, they usually won't even bother trying to read it. I certainly didn't when I was a reporter. In practice, this means that a research-oriented news release should be between 350 and 500 words long. Occasionally, a release might be slightly longer than 500 words, but that should be the exception, not the rule.

But if you're writing a blog post, I don't think word count matters. There is a school of thought that writing for the Web needs to be short. I don't agree. Writing, for the Web or anywhere else, needs to be *good*. If your piece is not interesting, people won't read it, regardless of length. But if your writing is engaging, people will read it no matter how long the article is. So don't focus on the word count. Focus instead on making sure you provide all the essential details while still writing a story that people *will actually want to read*.

JARGON AND TECHNICAL LANGUAGE

Words have specific definitions, yet people often use one word when they mean another—and this weakens many attempts to communicate about scientific topics. There are two issues here: first, make sure you use technical terms correctly; second, define those terms. This second point is significant, because even if you're using a technical term correctly, you should not assume that your reader will understand precisely what you mean. This is particularly true for scientific terms that people *think* they know and understand. For example, in a biological context, the word "population" refers to a group of individuals of a species living in a particular area; it does *not* refer to all of the members of a species. So if a *population* reaches zero individuals, the word "extinction" does not apply. It is only when a *species* reaches zero individuals that it is extinct. These are incredibly important distinctions, but they are ones that a nonexpert reader would probably not pick up on.

So try to keep jargon to a minimum, and explain any jargon you do use. Jargon is not inherently bad—it allows experts to convey large amounts of information in concise terms—but it can scare off or mislead readers, particularly if you use it early in a story.

FACT-CHECKING AND EDITING

Once you've written your blog post or news release, you need to make sure you didn't make any mistakes in either the science or the writing itself. For science PIOs, editing should be a two-step process: step one is to get the relevant researchers to go over it, and step two is to get another writer to read it. Share the draft release or post with the researchers you interviewed for the piece. Ask the researchers to track their changes so they are easy to spot. Also, if you're sharing the draft with more than one researcher, have them collate their corrections into one document and send you a single set of revisions. This allows you to avoid potentially competing changes, which can happen if you get revisions from multiple researchers in separate drafts.

When I send researchers a draft release, I ask them to "help me get it right." I ask them to point out anything that is technically incorrect. I also want to know if there is anything in the release that makes them uncomfortable: if they think I've overemphasized anything, placed the research out of context, or didn't give enough credit to research collaborators. This is important, because I want the researchers to feel that I'm representing their work accurately.

However, I do not allow the researchers to substantially rework my writing. Occasionally a researcher will rewrite a release so that it resembles a journal article abstract, packing it with technical language and inaccessible jargon. If that happens, remind them that you are not writing for an audience of their peers and that a news release or blog post is aimed at a non-expert audience. It can also be helpful to remind researchers that the goal of your release or post is to drive people to the research paper, which contains all of the technical details. You should always be polite when interacting with researchers, but it is important to remember that you are the expert when it comes to writing for non-expert audiences. You don't tell researchers how to design an experiment, and they shouldn't tell you how to craft a lede. When they are reviewing the piece, their job is to make sure the language is accurate, not to choose precisely which words you should use.

After you've gotten approval from the relevant researchers, you'll need to go through a second round of review. Ideally, you would get another professional writer to edit the work. You want to make sure that the release or blog post makes sense and is easily understood by a non-expert audience. You also want the editor to catch any grammatical or spelling errors. At NC State, we make sure that everything we publish is read by at least two other writers in order to minimize the possibility of any writing errors. If you work in a small organization, there may not be many writers who can edit your piece, but if there is any way to get another writer to read over your work, you should do so.

Finally, as I noted at the beginning of this chapter, some organizations—particularly government institutions—can require additional levels of review to account for political or other considerations. Find out what editorial controls exist within your organization, and make sure you comply with them. Depending on the organization, this process of obtaining additional reviews can be fairly time-consuming, so be sure to give yourself as much lead time as possible.

EDITORIALS AND OPINION PIECES

In addition to writing news releases and blog posts, science PIOs may occasionally be asked to help researchers or administrators write opinion pieces or editorials, or to ghost-write pieces on their behalf. Below are some basic guidelines for writing op-ed pieces. I share these with researchers and administrators who are writing their own op-eds, but the same rules apply to PIOs who are writing on behalf of their institutions. (This section is on writing an op-ed. I discuss how to pitch an op-ed in the next chapter.)

1. Make sure your op-ed piece is the appropriate length. Every news outlet has a word limit on op-ed pieces.
2. Write clearly and for a lay audience. It can be a tough balance: you don't want to assume that the reader is familiar with whatever issue you're writing about, but you also don't want to talk down to your audience.
3. Be sure you have a specific message that you are trying to convey. An opinion piece should provide perspective on and insight into an issue, rather than being a comprehensive overview of the issue.
4. State your opinion or make your argument right away. Ideally you will provide a synopsis of your viewpoint in the first sentence, and certainly in the first paragraph. Clearly laying out facts is great for supporting an argument, but you need to make the argument first and then back it up. (This is, after all, an opinion piece.)
5. Do not submit a rant about an article that has already been written. Save that for the "letter to the editor" section. Or, better yet, don't write angry rants at all.
6. Finally, remember to break up your sentences and paragraphs. Run-on sentences and endless paragraphs remove any rhythm from what you are writing. This point may seem trivial, but it's not.

KEY POINTS

- Know what you want to accomplish with your piece, and which audiences you want to reach, before you start writing.
- Determine whether you're writing a news release or a blog post.
- Write a good lede. You need to show the reader why they should keep reading.
- No matter what you're writing, you must be honest with your readers.
- If you're writing a news release, keep it short (and don't forget to acknowledge any funding agencies that supported the research).
- If you're writing a blog post, don't be afraid to experiment with different styles of storytelling.
- If you use technical terms, make sure to use them correctly and to define the terms for your readers.
- Get the relevant researchers to fact-check your writing.
- Get another writer to edit your work.
- If you're writing an op-ed, be sure to make a clear argument and support it. Also, make sure you comply with the relevant news outlet's word-count requirements.

PITCHING STORIES

It's not enough to write good science stories yourself; you have to be able to get reporters to tell stories for you. This is called "pitching a story," and it is an essential skill for science PIOs. Let's be clear about what good pitching is. It is *not* about convincing reporters that a mediocre story is actually a good one. That won't work. In a good pitch, the PIO serves as a liaison connecting reporters with stories that they'll actually want to tell.

This chapter will explain how to issue a news release, how to identify which media outlets you want to reach, how to work effectively with reporters, and how to use news services such as EurekAlert!

MASS DISTRIBUTION OF NEWS RELEASES

Science reporters and editors receive hundreds of news releases via e-mail every day. It is impossible for them to open all of these e-mail messages, much less read them. This is why it is important to have some perspective when planning to send out a news release. Unless a reporter has specifically asked to be added to your mailing list (which does happen), your news release is an unsolicited e-mail. In other words, it's spam. But that doesn't mean sending out news releases is a waste of time; it just means you need to manage your expectations.

Finding E-mail Addresses and Compiling Mailing Lists

Before sending out a mass news release via e-mail, you need to create mailing lists. These are collections of names and e-mail addresses for reporters. In order to target relevant reporters, you need to create multiple mailing lists. Each mailing list should consist of reporters who share a common characteristic. For example, you could have a mailing list for local reporters, another for science reporters in your geographic region, and one for national science reporters. The more specific your mailing list is, the more likely it is that reporters will actually read the news release.

Let's say you have a mailing list of national science reporters that includes reporters who focus on life sciences as well as those who cover space exploration. If you send out a news release about a biology study, the space exploration reporters will almost certainly not read it. Instead, they may mark your e-mail as spam, which will prevent them from seeing any future news release you might send. To avoid this problem, you should take the time to make separate mailing lists of national life science reporters, national space exploration reporters, and so on.

There are a lot of tools and services that can help you create and curate these mailing lists. Companies such as Vocus and BurrellesLuce provide searchable databases of reporters and news outlets that you can use to compile mailing lists. Some of these companies also allow you to send your news releases directly through their sites. The downside is that these services can be expensive. If your institution has sufficient resources, using one of these companies can be a good investment, and it's worth checking to find out whether your organization already has an account with one of these service providers. However, you can assemble a good mailing list without using one of these services. It just takes a little work. Begin by searching for recent news stories that are relevant to the mailing list you're trying to create. For example, if you're compiling a space exploration mailing list, search for news stories about NASA, Mars, or the International Space Station. Keep a file of the reporters who wrote the relevant stories. Then search for these reporters' e-mail addresses.

Some news outlets, such as the *Wall Street Journal* and the *Los Angeles Times*, usually include contact information for reporters at the bottom of their stories. Others consistently follow a formula for their reporters' e-mail addresses, such as the Associated Press uses, which is usually the initial of the first name, followed by the last name, @ap.org (e.g., Matt Shipman would be mshipman@ap.org). But some outlets, including the *New York Times*, have customized e-mail addresses for their reporters that don't follow a consistent pattern. If you type likely e-mail addresses into a search engine, you can usually find one that works. So you might search for matt.shipman@nytimes.com, shipmanm@nytimes.com, shipman@nytimes.com, matts@nytimes.com, and so on. If reporters have ever had their work e-mail address listed publicly online, you'll find it.

Once you've collected a good list of reporters and e-mail addresses, you have your mailing list. But if you've put this much time and effort into compiling a good list of reporters, you may want to save it for customized pitches (more on this topic later) because, in my experience, sending out mass news releases is only consistently effective at reaching one audience: small newspapers. Small newspapers usually have small staffs and small budgets. As a result, they will often run news releases from regional institutions with only minor changes, frequently listing the author as "staff reports." The term for this practice of repackaging news releases as stories from news outlets is "churnalism." Small regional newspapers located near your institution are also more likely to be interested in announcements of grants, events, partnerships, and new hires, because all of these things could have a bearing on the local economy. For these reasons, it makes sense to compile a mailing list of small newspapers near your institution and to include them on most—if not all—of your news releases.

Pushing News Releases on Your Website

If you've written a news release, you're going to want to make it available on your institution's website. This is important because the news release should be consistent with your institution's image and goals, but there are also practical media relations applications as well. Specifically, making the release available online is essential to your targeted pitching efforts (of which more anon) and because it can help you reach a number of news sites that don't rely very heavily on reporters.

Small newspapers aren't the only news outlets that practice churnalism. There are a number of science news sites that get much of their content directly from news releases. While most of these sites (e.g., Phys.org or redOrbit.com) lack the cachet of more mainstream news outlets, they do reach a significant number of readers.

Many of these science news sites get their content from the RSS (rich site summary) feeds of websites that disseminate news releases. An RSS feed allows users to "subscribe" to a website and

Things to Include in Pitches and Online News Releases

Several things should be included in any online news release or in any pitch you make to a reporter:

- Provide direct contact information for the researcher, unless your employer expressly prevents you from doing so.
- Provide a link to the relevant journal article if it is available online (even if the paper is not open access).
- Let the reporter know about any multimedia resources you have available: photos, graphics, video, or audio. (More on multimedia resources in the next chapter.)

get automatic updates whenever the site posts new content (such as a recent news release). For this reason, it's important to make sure that your institution's news site has an RSS feed that users can subscribe to. By the same token, if your organization is launching a new website or news page, you should contact relevant news sites to let them know it exists.

I'd focus on notifying sites such as Phys.org and redOrbit.com, which use RSS feeds from research institutions, because it is unlikely that science reporters at larger news organizations (e.g., the *New York Times*) are going to be interested in a research institution's new website. But if your organization is large or has a particularly high profile (e.g., NASA), reporters at mainstream outlets would probably be interested to learn about a new website or news page. Similarly, if you know reporters at large news outlets who focus on the specific research areas that your institution specializes in, it's worth letting them know about a new site.

Press Release Distribution Sites

Another way to disseminate your news release to a large group of reporters is to use an online press release distribution service. There are a number of these services available, but I'll focus primarily on EurekAlert!, which is the most prominent service for the research community. EurekAlert! is run by the American Association for the Advancement of Science and essentially serves as an online bulletin board. Research institutions (e.g., universities, journals, government agencies) pay a fee to subscribe to EurekAlert!, which then allows the institutions' PIOs to post news releases or other information about recent or forthcoming research news. This news could be about grant awards, patents, or research findings, including embargoed journal articles (more on embargoes later in this chapter). When posting a news release about research findings to EurekAlert!, PIOs are asked to list relevant research areas, where the findings are being published or presented, and how the research was funded. EurekAlert! also allows PIOs to give reporters direct contact information for researchers and to attach relevant multimedia files to news releases. Anyone can have access to most of the material posted on EurekAlert! and search for news releases in specific fields. Science reporters and editors can also register with EurekAlert! to get access to embargoed news releases and any material that PIOs have marked as "for reporters only," such as researchers' contact information.

When it works, EurekAlert! is a win-win situation. PIOs can share information with the reporters they want to reach, and reporters can find out about science stories they may want to cover. Full disclosure: I've worked with EurekAlert! for years in my capacity as a university PIO, and I've found it easy to use. The user interface is great, and it gives me the ability to reach reporters whom I might not have known about otherwise. One problem with EurekAlert! is that many freelance reporters and science bloggers are unable to register with the service because EurekAlert! will only allow full-time journalists to register as reporters, barring the admission of writers who are also scientists, students, or PIOs. Given that a significant percentage of science reporters and bloggers are freelancers who also have other jobs (including me), this prevents a sizable chunk of the science news community from having full access to embargoed releases and other "reporter only" information posted on EurekAlert! This should not discourage you from using EurekAlert!—it is still a valuable and effective tool for promoting research news and well worth the price of a subscription. Yet it's important to know the limitations of any product or service, and EurekAlert! is no different.

Other fee-based, online news release distribution services, such as Newswise, can also be effective tools for disseminating news to reporters. And there are three other news distribution sites of note that focus primarily (or exclusively) on science. The first is Alpha-Galileo, which is very similar to EurekAlert! but is aimed primarily at European audiences. The second is a free news release distribution service offered by the American Astronomical Society, which focuses on news items of interest to its members. The third is Futurity, which focuses solely on university research, although Futurity can be used only by a handful of university "partners." If your institution is one of those partners, it's a great resource.

TARGETED PITCHING

Mass distribution of news releases can be useful, but it's not the most effective way to get reporters to take an interest in research findings, grant announcements, or any other news story. Targeted pitching—approaching reporters and editors on an individual basis—is a far more effective way to get results. But it's tricky. A good pitch makes reporters aware of research findings they'll be interested in, which can lead to good news stories, making all of the relevant parties happy. A bad pitch is spam that clogs a reporter's e-mail inbox and makes it likely he or she won't read your e-mails in the future. You've got to find the right reporters at the right publications and give them the right information if you want them to cover your institution's work. This section will give you some practical information about how to do that.

Know What You Want

In chapter 2, I said that before you start writing, you need to know what you want. The same thing is true of pitching: once you've written a news release, you need to know what you want to do with it. To answer that question, it's essential to know your institution's goals and key audiences. I'll use my employer as an example. I work at a public university, and since the university's goal is to reach a wide variety of audiences, I want to pitch my story to reporters who write for mainstream media outlets. Local and regional news outlets (e.g., the local newspaper or TV station) may not be very exciting to researchers, but they are often the best way to reach the potential students, alumni, legislators, and industry partners that public university administrators care about. National mainstream news outlets (e.g., *Popular Science* or the *New York Times*) are good for much the same reason.

Most PIOs are familiar with the relevant outlets and can find the right reporter at each outlet with a little bit of work. (See the section below on how to pitch to reporters without being annoying.) But to reach the outlets that are most likely to help a researcher meet his or her goals, you will probably need help from the researchers themselves. (I touched on this briefly in chapter 1.) I often ask researchers where they go, outside of the peer-reviewed literature, to find out what is happening in their field. I'm looking for news outlets that run stories written by reporters but that target a specialized audience. These news outlets are usually discipline-specific or industry-specific. Does the researcher belong to a professional society that publishes a newsletter or magazine? Are there websites or blogs that focus primarily on research news in the researcher's specific field? These are discipline-specific outlets. In materials science, you can find *Materials Today* magazine or the *MaterialsViews* blog. Both are run by large publishers (Elsevier and Wiley-VCH, respectively), but cover research regardless of what journal the research was published in. And both are read almost exclusively by materials scientists. Maybe only a few thousand people will see a story that runs in either outlet, but all of those people matter. They are likely the people with whom your researcher will want to collaborate—and they are almost definitely the people who may cite the research in their own papers.

Industry-specific outlets are more application-oriented. For example, if I'm responsible for promoting materials science research that may be relevant to the development of new semiconductors, I'll want to target news outlets that are read by the semiconductor industry. And there's a lot to choose from: *Semiconductor Today*, *Compound Semiconductor*, and the SEMI.org family of newsletters, to name a few. The audience for these publications include people in the private sector who may license new discoveries or fund additional research, as well as those at federal agencies who want to demonstrate the value of research to Congress. But whether a news outlet is industry-specific or discipline-specific, you may not even know it exists if you don't ask for input from the researchers you are working with. And you certainly aren't going to know offhand which outlets the researchers think are most important.

It's also worthwhile to pay attention to news outlets that are run directly by federal agencies. I pitch stories to the National Science Foundation's Science360 site, and they often choose to highlight the work I send them. These news outlets reach a relatively small but important audience. In my experience, the only thing that makes researchers happier than seeing their work highlighted by a federal funding agency is receiving a grant from a federal funding agency.

Once you've decided which news outlets you want to reach, you need to figure out which reporters you should contact.

Do Your Homework

Whether you're planning to pitch a story to a local newspaper, a mainstream media outlet, or a niche publication, take the time to find out which reporters cover the topic you're pitching. Read their stories and familiarize yourself with the publications they write for. For example, just because a reporter writes for *Scientific American* doesn't mean he or she covers every scientific issue. And nothing is a bigger waste of time—for you and the reporter—than pitching off-topic. If a reporter mainly covers astronomy and astrophysics, he or she is not going to want to write about new genetics research, no matter how interesting it is.

Figuring out a reporter's beat isn't always easy. Some reporters seem to write about anything and everything. In those cases, I will often simply e-mail reporters and explain that I'd like to give them a heads-up about some research they might be interested in, but I can't get a handle on their beat because it's so varied. Then I ask them what sort of pitches they'd be interested in. Most of the time, I don't get a response. (If it's a reporter I don't know, I assume he or she didn't even open my e-mail.) But sometimes reporters give me clear guidelines about the kinds of research they want to know about. That's enormously useful. They get decent leads from me, and I know I'm sending them things that they'll at least find interesting, if not always story-worthy.

In fact, I usually ask reporters for guidance on what they're interested in, but only after I've made it clear that I have some understanding of their beat. This often happens when a reporter I don't know contacts me for information about a news release—for example, after an editor assigned them to cover the research or they ran across it on EurekAlert! I might know that the reporter covers neuroscience, but neuroscience is a pretty big field. What are the reporter's particular areas of interest? That said, don't make it sound like you know nothing about what a reporter covers. That's a surefire way to get him or her to ignore you.

Lastly, it is never a good idea to pitch the same story to two reporters at the same news outlet. If they both decide to follow up on the story and don't find out about it until later, they'll be angry. Their news outlet is only going to publish one story about the research, which means that one of the two reporters is wasting his or her time. So if you want to build a good working relationship with reporters, do *not* double-dip.

Keep It Short

First of all, you'll want to pitch to reporters via e-mail. You should not call reporters on the phone to make your pitch. This is intrusive, annoying, and almost definitely a waste of their time and yours. If you've established a good working relationship with a reporter, you may be able to call him or her. But even then, don't do it very often.

However, as I noted earlier, reporters and editors get a *lot* of unsolicited e-mails. And no matter how good your intentions are, your pitch is one of those unsolicited e-mails. If you're writing a reporter you've worked with before, he or she may open your e-mail. If not, you're probably out of luck. But to improve your odds, you need to have a concise subject line that tells them why they should bother reading it. To hold their attention, keep your e-mail *short*. This can be challenging because research news is often complex and you don't want your pitch to be misleading. It can also be difficult, because you'll want to present reporters with a news package, letting them know about relevant contacts, multimedia resources, and anything else a reporter might need to follow up on the story. Keeping all that in mind, here's a good formula for constructing your e-mail pitches, with components listed in the order in which they should appear:

1. One or two sentences explaining what the work is and why it's important.
2. A link to additional information, such as a news release (this is why you post it online).
3. A link to the journal article itself, if available.
4. Direct contact information for the researchers, so reporters don't have to contact you again if they don't want to.
5. Links to related multimedia materials, such as images or video, if available.
6. A note to let them know you can provide a copy of the paper, et cetera, if the article is not open access.

Note: I know that some research institutions, such as federal agencies, do not provide direct contact information for the relevant researchers. Instead, they have all media contacts routed through their PIOs. Make sure you comply with the rules and regulations in place at your institution.

A quick rule of thumb about pitches: if you look at your e-mail and wonder whether it's short enough, it's not short enough. Here's an example of an e-mail pitch I sent to a life sciences reporter:

> Hey, Ed:
>
> Thought this might be of interest. Researchers have developed a *de facto* antibiotic "smart bomb" that uses CRISPR RNA to identify specific strains of bacteria and sever their DNA, eliminating the infection. This would allow us to kill "bad" bacteria without killing off "good" bacteria, and also offers a potential approach to treat infections by multi-drug resistant bacteria.
>
> Worth noting: this was done in a lab, so there are still a lot of questions about how to apply this in a clinical context. (Want to state that up front.) Still, pretty cool.
>
> More info here: http://news.ncsu.edu/releases/wms-beisel-crispr2014/
>
> Full paper is here (open access): http://mbio.asm.org/content/5/1/e00928-13
>
> You can contact researcher Chase Beisel directly at (919) 513-2429 or cbeisel@ncsu.edu. (He responds much more quickly to e-mail messages.)
>
> Please let me know if you'd like anything else from my end.

One other note: if you think that only one or two news outlets will be interested in a particular news item, you don't need to write a news release. Instead, pull together relevant materials, such as the journal article and the direct contact information for the researcher, and pitch to the reporter (or reporters) that you think will be interested. This is particularly true for news that will only be of interest to local reporters or to reporters at discipline-specific and industry-specific news outlets.

Be Relevant

Tailor your pitch. You may know that a reporter covers medical research, but you need to be aware of their focus within that field, and you should make clear how any new research is relevant to their specific beat. You also should take into account the angle that a reporter's news outlet is interested in. For example, if you're pitching medical research, newspapers will generally want to know how the findings will affect patients, business publications will be interested in potential commercial applications, and technology publications will focus on new medical devices. Again, this comes back to doing your homework. Crafting a good pitch takes time and effort, but it's worth it. Not only are you more likely to see the research covered in a news outlet, but you'll also cultivate good working relationships with reporters. That means they'll be more inclined to listen to your pitches in the future.

Coordinate with Other Institutions

If you're writing about research that was conducted by researchers at two or more institutions, try to coordinate with the PIOs at the other institutions. This way, each PIO can focus on a specific region or set of reporters. There are three advantages to this. First, a coordinated effort means that a reporter won't get multiple pitches on the same research, which can be annoying. Second, a coordinated effort can prevent double-dipping, mentioned above as a very good way to alienate reporters. Third, a good PIO builds working relationships with reporters, and a coordinated effort will allow PIOs at the relevant institutions to take advantage of those relationships. So if the PIO from Institution A has a good contact at the *New York Times*, let her be the one to make that pitch. If the PIO from Institution B has a good contact at *Popular Science* and *Wired*, let him make those pitches. Work together and play to your respective strengths.

Be Responsive

It's important to remember that pitching to a reporter is the beginning of the process, not the end. Your goal is to get reporters to cover your institution's research. This means that even if you've included direct contact information for the researchers and links to supporting material, reporters may contact you with additional questions. If a reporter does follow up and ask you for something—such as a copy of the journal article, related images, or a headshot of the researcher—take care of it as quickly as possible. If reporters ask for something you can't provide, let them know as quickly as possible so they have time to figure out what other options might work. You should always assume that reporters are working under a tight deadline and if you don't respond quickly, they may drop the story altogether. Even if they aren't under a rush deadline, they will appreciate a prompt response.

By the same token, if reporters have general questions about the work, feel free to answer. To be clear: Do *not* block their access to the researchers, and do *not* answer on behalf of the researchers. But if they want to know whether a finding is relevant to a specific research area, it's fair to say yes or no, as long as you qualify it. When this comes up, I'll say something like: "My understanding is that this [is/is not relevant to a specific area], but you really need to ask the researchers." I do this because sometimes the reporter is asking in order to get a broad sense of the subject before calling the researcher. Just make sure that the reporter has access to the researchers and knows that they have the final word. If the researchers don't want to talk to reporters, you shouldn't be pitching their work in the first place.

EMBARGOES

Embargoes are a fact of life for science PIOs and reporters. But a lot of people don't know what they are or why some institutions require them.

How Embargoes Work

In a science news context, an embargo is when a journal, researcher, or PIO gives reporters a copy of a journal article before it is published and bars those reporters from releasing any stories about the article until it has been published. Reporters are given advance notice of the findings and a preview copy of the relevant article so they have time to read it, conduct interviews, and generally do their homework before the journal comes out. In theory, this means that when the journal article does become publicly available, news outlets will simultaneously unveil well-reported news stories on the article's findings. For example, here's the language that the *Proceedings of the National Academy of Sciences* journal uses to explain the purpose of its embargo policy: "This policy is designed to provide news reporters an opportunity to write accurate news stories while ensuring that publicity does not appear prematurely. Through consistent implementation of our embargo policy, we work to maintain a fair and level playing field that gives no one reporter or organization an advantage over others."

Embargoes are particularly useful for news outlets such as newspapers and magazines. If a magazine is scheduled to come out on September 21, its editors will need to send the final version to the printers by, say, September 7 (if not earlier) so they can print and ship the issue. If an exciting new study is scheduled to be published in a research journal on September 20, the magazine would not be able to cover the study unless the reporters and editors had gotten an embargoed copy. If the magazine's reporters *do* have an embargoed copy, they can write about it and send the article to the printers, knowing that the study will be public knowledge by the time the magazine hits streets on September 21.

Printing and shipping are no longer issues for many news organizations—they can make their stories available online at any time. But this only highlights another argument in favor of embargoes: the 24/7 news cycle. With the arrival of around-the-clock news channels and websites, a greater premium has been placed on being the first to report a story. In many instances, it seems that news outlets are far more concerned with being first than with being best—not to mention doing a decent job of reporting a story. An embargo addresses this problem by giving reporters the chance to do a thorough job of covering the story ahead of time instead of scrambling to pull together a story on short notice. Thus, embargoes provide real value.

How to Handle Embargoed Material

If you're responsible for determining whether research findings should be embargoed, there are two key points to bear in mind. First, make sure the relevant findings are not already public. You cannot embargo material that is already available online or elsewhere. Second, if you're going to go through the effort of establishing an embargo, you should make sure that the embargo gives reporters enough time to read the article and write a news story. However, PIOs are rarely in control of establishing embargoes, which are usually put in place by journals. The journal's publishing company will set a date and time when a research article is scheduled to be made public, and they'll share that information with the researchers who wrote the article and (sometimes) with the PIOs at the relevant research institutions.

Now I'm going to share one of my pet peeves with you: If you choose to pitch embargoed material to reporters, *do not* disseminate the material under the assumption that a reporter is automatically obligated to follow the terms of the embargo. For example, a PIO should never send a reporter an unsolicited e-mail containing a prepublication copy of a journal article and a note saying, "This is embargoed until [date]." The reporter didn't ask for the paper and therefore didn't agree to the terms of the embargo prior to receiving it. This means that the reporter would be well within his or her rights to write about the paper and ignore the embargo. If you pitch a story about an embargoed paper, wait until the reporter has agreed to the terms of the embargo before sending them the article. The only exception to this is if you are putting embargoed material on EurekAlert! or a similar news distribution service. If reporters have registered with EurekAlert!, then they have agreed to abide by the terms associated with any embargoed material on the site.

PRESS CONFERENCES

Press conferences bring together large groups of reporters and all of the major players in a news story. They enable researchers and administrators to unveil new information, and they give reporters an opportunity to ask questions of relevant sources in a public forum. Press conferences are usually held to announce particularly important research findings, such as high-profile medical advances, grants that are so large that they make a substantial impact on a research institution, or events such as breaking ground on a new research facility. The advantage of a press conference is that it gives an institution the ability to control precisely when new information gets out and allows all of the relevant parties at the institution to speak with one voice at one event.

That said, I am not a fan of press conferences. In most instances, press conferences are attended only by local reporters. If your research institution is located in a major media hub—such as New York, London, Los Angeles, or Atlanta—those local reporters may include representatives from media outlets with a national audience. However, if you are located in a smaller market—such as Raleigh, North Carolina, or Madison, Wisconsin—a press conference will likely be attended solely by reporters from local TV affiliates, newspapers, and radio stations. Local news coverage is important, but a press conference will not necessarily achieve anything that you couldn't have accomplished by pitching to local news outlets individually. Similarly, you can just as easily control the timing of when new information is released by issuing an embargoed news release.

If you're planning to hold a news conference, there are a number of things to bear in mind:

- Issue a short news release inviting reporters to the event, saying who, what, when, where, and as much of "why" as you can (these kinds of news releases are called media advisories).
- Time the media advisory (an event-oriented news release) to go out one or two days before the event; otherwise, the event may get lost on reporters' calendars. A sample media advisory is included in appendix D.
- Send individual pitches to any reporters you particularly want to attend the event.
- Make sure anyone speaking from your institution has media training and knows what they're going to say (more on this in chapter 5).
- If possible, block out time for the key players at your institution to talk to reporters individually before or after the event.
- Ensure that reporters will have everything they need: a mult box for camera operators to get audio of the speakers, fact sheets or other relevant supplemental materials, and so on.

BECOMING PART OF SOMEONE ELSE'S STORY

Getting reporters to write about the work that your institution's researchers are doing is not the only way to raise your institution's public profile. You can also find stories that reporters are already telling and identify ways to make your institution part of that story. There are a number of approaches you can take to insert your institution into an existing news story, but all of them rely on your ability to provide insight and analysis for the existing story. You need to be able to give reporters information that helps them place the story in context or gives them a fresh angle on the story.

Some news stories are predictable. For example, every winter, reporters will write articles about Thanksgiving and religious holidays; every two years, writers are covering either the Summer or Winter Olympics. Look for ways to become part of those stories. Medical or food safety researchers might be able to talk about foodborne illness and how to keep your family safe during Thanksgiving. Similarly, the Olympic Games give doctors the opportunity to talk about athletic injuries, while physicists can discuss the physics of curling or the use of leverage during the pole vault.

Other news stories are less predictable, but you can usually tell if they are going to be in the public eye for an extended period of time. The 2010 Deepwater Horizon oil spill in the Gulf of Mexico was clearly going to be the subject of news stories for months or even years. The oil spill also created opportunities for a wide variety of researchers to work with reporters. Reporters wanted to know about the spill's effects on local ecosystems, marine animals, and public health. They also wanted to know how oil rigs worked, what might have gone wrong, and what sort of technologies would be used to clean up the spill. Biologists, veterinarians, medical doctors, and engineers were among the many researchers who were asked to comment on various aspects of the spill and its aftermath.

Regardless of whether a news story is predictable, there are things you can do to be prepared and to take advantage of news opportunities. First, you need to be familiar with the researchers and their specialties at your institution. If you have an engineering researcher who is an expert on how oil rigs are designed and operated, this won't do you any good if you don't know he or she exists. Know researchers' strengths—but know their weaknesses as well. If you have an oil rig engineer who is unable or unwilling to use language that can be understood by non-experts, you don't want to recommend that expert to TV reporters. Print reporters might be able to take the time to work with the expert for their stories, but most TV reporters are looking for quotable clips, and your expert won't fit the bill.

Second, obviously, is that you have to pay attention to the news. If you're not tracking the daily news, you won't know what stories are out there, much less how they may relate to your institution. I also ask researchers I work with to let me know if they run across stories that they want to be a part of, such as if they hear a story on NPR while driving in to work and they think, "Gosh, the reporter really missed an important point there." I might not have heard the story, or I might not have seen how it was relevant to a researcher's specific field of interest.

Third, if you have an expert who is a good fit for a news story and has the communication skills to work with news media, make sure that the expert is willing and able to talk to reporters. Researchers usually don't like being surprised by reporters, and you won't know ahead of time if the researcher is even available—he or she may be on vacation, at a conference, or otherwise out of reach. Lastly, reaching out to the researcher in advance lets you find out whether the researcher has any potential conflicts of interest. For example, if your oil rig engineer was a consultant for a company associated with the Deepwater Horizon spill, you would have to tell reporters this when putting the engineer forward as an expert. Reporters would likely still want to interview your engineer (probably even *more* so), but they could no longer view your engineer as an impartial observer because the expert has a stake in the game.

Finally, you need to let reporters know that your experts exist. One way to do this is to contact reporters who have already written about the relevant news event and tell them that your expert can provide additional insights for any follow-up stories. This sort of customized pitch can be very effective, particularly for news stories that will definitely be in the public eye for an extended period of time.

But you can also create a specialized news release to be broadcast (and pitched) to a large number of reporters. This type of news release is called a "news tip" and can focus on one or more researchers who have expertise relevant to a news story. A news tip should include a one-paragraph summary of the relevant news item and state that your institution has an expert or experts who can offer insight and analysis for that news item. Then list each expert's name, contact information, expertise, and how that expertise is relevant to the news story. You disseminate and pitch a news tip the same way you would a more conventional news release.

ProfNet and Expert Lists

There are two other ways to make your researchers part of an external news story: ProfNet and online expert lists. ProfNet is an online news service that connects reporters with experts. Institutions pay a fee to get access to ProfNet, where reporters post requests for experts in specific fields. For example, a reporter might post a request asking to speak to a psychologist about dissociative identity disorder. A reporter can list her name, publication, and contact information, allowing PIOs to contact her directly, or a reporter can choose to remain anonymous, in which case PIOs respond via the ProfNet site. Either way, a PIO can send the reporter a short note with the name and contact information of a researcher who has the expertise that the reporter is looking for. If you're willing to make the time to peruse ProfNet on a regular basis, it can be a very good investment.

Some institutions also choose to create searchable online lists of experts that reporters can use to find experts on a variety of subjects. These can be incredibly useful tools for both reporters (who can go straight to the expert without dealing with PIOs) and for PIOs (who may have trouble keeping track of who all their experts are). However, online expert lists can be problematic in several ways. First, creating an online expert list is an enormous amount of work. You not only need to inventory the expertise of all of your institution's researchers (or at least the ones who are willing to talk to reporters), but you also need to plug all of that information into a database and create a searchable website to make that information accessible. This can be both expensive and time-consuming, particularly for large institutions.

Second, you will need to maintain that database to make sure it is current. If there is a significant amount of employee turnover at your institution, this can be difficult; but if your expert database isn't up-to-date, it isn't useful. Finally, making a comprehensive, up-to-date expert list isn't enough. You need to get reporters to use it. If you work for a high-profile research institution, this may be relatively easy, but there is often little incentive for reporters to seek out experts at smaller institutions. At any institution, you will need to create and maintain awareness of the expert list by reaching out to reporters. This can be a tedious and time-consuming task.

Placing Op-Eds

Here are some basic things to consider when trying to place opinion or editorial pieces (op-eds) in news outlets (tips on writing op-eds are given in chapter 2):

- Do your homework and make sure the op-ed fits within the publication's designated word count. You don't want to waste their time or yours.
- Make sure you're submitting the op-ed to the correct person or e-mail account at the relevant news outlet.
- Specify in your e-mail that you are submitting the op-ed exclusively to that news outlet.
- Remember that you promised exclusivity, so do not submit an op-ed to more than one news outlet at a time. If you haven't gotten a response from a news outlet within one or two weeks, feel free to submit it elsewhere.
- Briefly state who wrote the op-ed and why they are qualified to write about the subject (this statement should not be more than one to two sentences long).

FREEDOM OF INFORMATION ACT AND PUBLIC RECORDS

If your institution is a state or federal entity, reporters can request access to much of your institution's information. This can be tricky legal territory, so you should notify your legal counsel whenever you receive a public records request. What I can provide here is a very broad overview of some of the issues at play.

The Freedom of Information Act (FOIA) is a federal law that applies to federal entities, requiring them to give any member of the public access to a wide variety of information. States have similar laws for state agencies. However, there are a number of exceptions to these laws, and state and federal laws are often very different in regard to which kinds of information are confidential and which are matters of public record. Some types of information are, to the best of my knowledge, universally considered to be confidential. For example, I am not aware of any public information laws that require institutions to make employee social security numbers available to the public.

Legal requirements for disclosure of research-specific information can be significantly more complicated, and laws governing the release of such information vary widely from jurisdiction to jurisdiction. Some states have laws exempting research information from public records requests, giving research institutions significant protection from having to make research notes and records public. Other states offer limited protection or no protection at all against public requests for research information. For example, would unpublished research findings be exempt from public records requests because they may constitute technical information with commercial value? This will vary according to the relevant statutes and case law in your jurisdiction. That is why it is a good idea to work with legal counsel.

Another good reason to consult with legal counsel is to help you avoid violating privacy laws when complying with public information laws. For instance, if you make information available about students who were engaged in a specific research project, you might risk exposing your institution to legal action for violating the students' rights to privacy. Again, talk to your lawyer.

Not only do you need to comply with public information laws; it is also in your best interest to do so quickly. If you are able to produce the relevant public records in a reasonable amount of time, reporters will likely take the records at face value. But if you drag your feet and don't produce the materials for weeks (or longer), they will assume you're hiding something, even if you aren't. Remember: You want to build and maintain a good working relationship with reporters.

It's important to familiarize yourself with your institution's established procedures for handling public information requests. If your institution does not have a process in place, you should push for it to establish one.

KEY POINTS

- Know which news outlets you want to reach.
- Compile good mailing lists.
- Do your homework, and pitch to the right reporters.
- Keep your pitch short.
- Include contact information, a link to the paper, and multimedia resources (if possible).
- Be responsive to reporters' requests for additional information.
- Don't embargo information that's already public.
- Don't send embargoed materials to anyone who hasn't agreed to the embargo.
- Get photos or other multimedia if you can.
- Track daily news, and look for story subjects where your institution has relevant expertise.
- Know who your experts are and how they might be able to contribute to ongoing news stories.
- Follow your institution's procedures to comply promptly with FOIA or other relevant public records laws, and work with legal counsel when appropriate.

ILLUSTRATING STORIES WITH MULTIMEDIA

Having a great story, a well-written news release, and a good pitch are important if you want reporters to cover your research. But there are a lot of good stories out there, so set your institution's research apart by providing reporters with supplemental multimedia materials: video, audio, photographs, or illustrations. This chapter is not intended to be a tutorial on how to shoot video, record audio, or otherwise create and edit multimedia materials. There are entire books devoted to every aspect of multimedia production. Instead, I want to offer a brief overview of the importance of multimedia as well as some suggestions for how to find relevant resources.

WHY MULTIMEDIA MATTERS

A good image can make your research news stand out. I wrote a news release once about a research team that had created a nanoscale, flowerlike structure out of a semiconductor material. The research was extremely practical because the structures had a very large surface area and held promise for use in applications such as energy storage devices. However, what got the attention of reporters was the image that went with the news release. The image, created via advanced microscopy techniques, was beautiful; the structure resembled an ornate flower, such as a chrysanthemum blossom. As a result, news outlets that might not have normally reported on nanotechnology took an interest in the story. Once reporters found that the pretty picture was accompanied by robust science with practical applications, many of those news outlets decided to cover the story.

But images can do more than simply get attention. They can also help people understand how the research works or why it is important. For example, photographs can be much more effective than words alone at illustrating the difference between a healthy plant and a plant afflicted with a specific disease. Similarly, audio recordings are much better than written descriptions at conveying differences in the mating calls of frog species. The same principle holds true for video.

Multimedia is also important if you're planning to disseminate news via social media outlets. For YouTube, of course, video is essential. But good images are also crucial if you want people to click on links in Facebook, Google+, or other social media platforms. Text alone is rarely sufficient to garner attention in a user's crowded news feed.

Now that you know some of the reasons why multimedia elements are important, let's talk about where you get your multimedia resources.

RESEARCH IMAGES

The first people to go to about multimedia resources are the researchers themselves. The researchers may have used microscopes as part of their study, taken photographs while doing field work, or recorded video of a laboratory experiment. Whether they were capturing multimedia as part of their research or simply recording a project they were excited about, these multimedia resources can be enormously valuable. Some researchers I've worked with make videos of their experiments and place the videos online, where I can link to them in our news releases. At least four of those videos have garnered more than 100,000 views each on YouTube. These are not professional-quality videos, but they do offer a firsthand look at research that is both interesting and visually appealing. (Some of the relevant research deals with liquid metal, which looks cool.) It's also worth noting that I encourage the researchers to keep the videos *short*. The longer a video is, the less likely it is to hold someone's attention.

More often, researchers are able to provide me with photos and microscopy images that they've taken as part of their research. These are also useful. For example, images of microneedles or nanoscale scaffolds used in biomedical research can be very visually interesting. You never know what multimedia resources researchers have unless you ask. They may have something wonderful that will make your news package significantly more appealing.

Lastly, find out whether the researchers' multimedia resources were used in a journal article. If so, the journal's publisher probably owns copyright to the multimedia. This shouldn't be a problem, but it does mean that you must contact the publisher and get approval to use the image. Every time I have asked a publisher for permission to use an image or other multimedia resource, they have said yes, as long as I have given appropriate credit to the publisher. The "appropriate credit" in this context is whatever acknowledgment the publisher tells me to use.

WORK WITH PROFESSIONALS

If researchers don't have multimedia resources for you, try to work with professionals to create some. A professional photographer will take better photos than an amateur, a professional graphic designer will create better illustrations, and the same holds true for other multimedia formats. Many research institutions, such as federal agencies and large universities, employ photographers, videographers, and graphic designers. PIOs should establish good lines of communication with these multimedia experts and take advantage of their expertise when possible.

There are a couple things to bear in mind when working with multimedia pros. First, make sure you give them enough lead time. A photographer may have a crowded calendar and need days or weeks of advance notice to schedule a shoot with relevant researchers. Videographers and graphic designers usually need even more lead time to create a product that you'll want to use. If you have advance notice of exciting research findings and you want to incorporate multimedia elements into your press package, bring the multimedia professionals into the process as early as possible.

Second, make sure the multimedia experts have a clear idea of what you want or need. For example, if you want a graphic designer to create artwork that illustrates a cellular process or a piece of nanotechnology, you need to work with both the researchers and the artist to make sure that the artist understands what he or she is supposed to be producing. You don't want an artist to spend a significant amount of time on a project only to find that the illustration is inaccurate or misleading about how the research works. If you're working with videographers, it might be worth asking them to get some good "b-roll" footage—that is, supplemental footage that shows the lab space, machines in operation, researchers at work, and so on. Many news outlets are increasingly open to the idea of combining their own reporting with raw footage from an institution to edit together their own videos. This approach is particularly useful for research that is visually compelling.

STOCK PHOTOGRAPHY AND OTHER IMAGE RESOURCES

Unfortunately, a lot of research institutions do not have multimedia professionals on staff. Even institutions that do employ photographers and other multimedia pros aren't always able to meet a PIO's multimedia needs, whether because of time constraints or other reasons. And while video and audio resources may be a luxury, graphic images are virtually a necessity for promoting research news. Luckily, PIOs have access to a wide variety of online image resources. I'll talk about various types of online image libraries, but it is important to note that you should always provide an attribution for any image you use (i.e., list the name of the photographer). It is also good practice to provide a link to the source of the image.

One option is to search the Web for relevant images. Photo-sharing sites such as Flickr can be particularly useful resources. For instance, if you're preparing to promote research about coastal habitats or a certain species of bird, you can search Flickr or similar sites for relevant photographs. If you find an image that is appropriate, you can then contact the photographer and request permission to use the image. If you don't get permission, don't use it. There's also a Creative Commons (CC) search function within Flickr that allows you to find images that have CC licenses, meaning you don't even have to get permission as long as you follow the license's requirements (e.g., appropriate credit).

Another option is to look for images on government websites, such as the Images from the History of Medicine website maintained by the U.S. National Library of Medicine. These sites offer a wide array of images available for use by the public, as long as you agree to the terms of use associated with each image. These sites also provide information about attribution and copyright restrictions regarding their images. Other useful government image sites include the U.S. Geological Survey Photographic Library, the Public

Health Image Library (maintained by the U.S. Centers for Disease Control and Prevention), the National Oceanic and Atmospheric Administration Photo Library, the National Science Foundation Multimedia Gallery, and the NASA multimedia site. (There is a list of online image libraries in appendix A.)

There are also a number of other websites that serve as free "stock photography" libraries, providing users with access to an array of photographs that can be used for specific purposes. Examples of these sites include Wikimedia Commons (which also offers stock videos) and Stock.XCHNG. However, make sure to abide by the attribution and permission requirements associated with each image. While all of these websites provide attribution information with each image, the sites vary in how they provide usage or permissions information. For example, Wikimedia Commons includes attribution and permissions data on each image's webpage. Other sites use a blanket permissions guide covering all of the images on the site, while still others require users to click on a link to each photographer's account page to learn what uses are permitted.

Lastly, there are also commercial stock photography sites, such as iStockphoto and Big Stock Photo, which require you to pay for stock images. While these sites do charge you for the images, they are otherwise very similar to free stock photography sites. The primary advantage of commercial stock photography sites is that they offer a much broader range of images, and the images are often of higher quality. I usually turn to commercial stock photography sites only after having tried—and failed—to find appropriate images on free sites.

DOING IT YOURSELF

This book is not intended to provide an in-depth tutorial on how to become a competent multimedia professional, but I do want to talk briefly about the potential that exists for PIOs to create multimedia content. (There is also a list of multimedia tutorial resources in appendix A.) Taking photographs requires equipment and photo-editing software (to crop your photos, if nothing else). But even small institutions often have a sufficient budget to buy a camera and a photo-editing program. Smartphones are also capable of taking, and editing, high-quality images. If you're interested in taking your own photographs, do your homework: familiarize yourself with your equipment and software; take advantage of online tutorials; talk to professionals or accomplished amateurs; and practice—a lot. You may never be able to take photos that rival the professionals, but with practice most people can produce acceptable photographs.

It is much more difficult to produce good videos than it is to take good photographs. In addition to relevant equipment and software, you need to have a steady hand (or tripod), the ability to record good audio, and the skills to edit the footage into a compelling, visual story. The equipment and software can actually be the easy part. Smartphones allow users to capture video and audio of acceptable quality, and there are a number of video-editing apps as well. Again, the key is to do your homework: familiarize yourself with your equipment and software; take advantage of online tutorials; talk to professionals or accomplished amateurs; and practice. Sound familiar?

If the subject matter is compelling, an "acceptable"-quality video is still extremely useful. In other words, if you're doing a story on research that *looks* really cool, you should definitely try to make a video. Even if the video isn't completely polished and professional, it can capture the imagination of reporters (and anyone else) and may give reporters an idea of the visual potential of a story. If they can see the potential in your video, they may want to come shoot their own professional footage.

Audio is the trickiest of all. Short snippets can be useful, such as when I used very short recordings of frog mating calls in a blog post about that topic. However, I wouldn't recommend investing time and resources in longer recordings or podcasts. I've talked with several reporters who work in audio about this, and they agreed with me. "I don't think I'd suggest a podcast," said Rose Eveleth, a freelance reporter and cofounder of Science Studio. "It takes a long time to make a good podcast. You can make a really bad podcast pretty easily. But what's the point of that? Even a bad podcast will take you several hours a week, and what you'd be getting out of it isn't clear."

COPYRIGHT, FAIR USE, AND RELEASE AGREEMENTS

Whether you're using multimedia resources that you've created or images or videos from elsewhere, make sure you're on solid legal footing. Whenever you're using images that don't belong to your institution, it is important to make sure you're in compliance with copyright law. If you have any questions about defining copyright or determining whether you may be violating copyright law, you should contact your institution's legal counsel. Here is a very broad overview: a copyright is the legal control that photographers, artists, writers, musicians, and other creators of original material have over the material they've created. For example, copyright law is what prevents people from legally selling pirated copies of CDs and DVDs. However, under the doctrine of "fair use," people are allowed to make use of copyrighted materials under certain circumstances. News reporting and nonprofit educational uses are often considered fair use and therefore not violations of copyright law. Commercial uses, on the other hand, are usually not covered by fair use.

Similarly, if you're taking photographs or making videos, you may need release agreements for the models or properties appearing in the photos or videos. These agreements indicate that you had consent from individuals or the representatives of locations to photograph or video them. Release agreements protect you and your institution from lawsuits claiming an invasion of privacy or defamation. As with copyright and fair use, the right of privacy has exceptions. For example, photos and videos used in news reporting are often exempted. You are also likely safe if you're taking photos or video of property owned by your institution. However, commercial uses of photography or videos are often not considered exempt from privacy law.

Columbia University has developed a good overview of copyright and fair use that may be helpful if you are unfamiliar with fair use guidelines. Columbia's Fair Use Checklist is available online at http://copyright.columbia.edu/copyright/fair-use/fair-use-checklist/. The American Society of Media Photographers offers a good overview of release agreements and relevant privacy law, available online at http://asmp.org/tutorials/property-and-model-releases.html#.

Make sure you are in compliance with whatever permissions and agreements apply when creating or using multimedia. And this is worth repeating: If you have any questions about whether your use of an image, video, or other copyrighted material may violate copyright or privacy laws, contact your institution's legal counsel.

KEY POINTS

- Multimedia isn't just window dressing; it's important.
- See if your researchers have images or video you can use.
- Work with professionals whenever possible.
- Make use of the many stock photography sites out there.
- If you're going to produce your own multimedia content, do your homework and practice, practice, practice.
- Don't use low-quality multimedia—or any multimedia that you don't have permission to use.
- Make sure you attribute multimedia resources if they came from other people—and that you are complying with copyright law.

5

GETTING SCIENTISTS TO TELL THEIR STORIES

If reporters are interested in your research news, they are going to want to talk with the people who actually did the work. This chapter will explain how to work with researchers so that they will be comfortable and effective at discussing their research during interviews with print, online, radio, and TV reporters. It will also address the role of PIOs during interviews.

MEDIA TRAINING FOR PRINT AND ONLINE INTERVIEWS

Scientists are often nervous about being interviewed by reporters. This is usually because they are worried that reporters will misrepresent their work or take their quotes out of context. Unfortunately, there is no foolproof way to ensure that reporters will get everything right. However, there are things that scientists can do to help explain their research effectively and to significantly improve the odds that their work is presented accurately. As a PIO, part of your job is to train your researchers in how to interact with reporters. This section is essentially a training module you can use to prepare researchers for print and online interviews.

Take Your Time, Do Your Homework

If researchers get a call from a reporter who wants to interview them about their work, they shouldn't rush into the interview. Instead, instruct researchers to ask the reporter a few questions first: Who is the reporter? Which news outlet is he or she writing for? What, specifically, does the reporter want to know about? For example, if the reporter is calling about a specific paper, are they interested in the science for its own sake? Or are they primarily interested in potential medical or commercial applications of that science? In short, it makes sense to get an idea of what the interview will focus on.

Once researchers know who the reporter is and (broadly speaking) what the reporter wants, they should get the reporter's contact information and tell the reporter they will call back in ten minutes. Most reporters are perpetually on deadline, but if they want to talk with a researcher they can usually spare ten minutes. This gives researchers time to prepare.

Researchers should use that ten minutes to do some quick homework on the reporter and the news outlet, if they are not already familiar with either of them. If a Google search turns up information that makes researchers nervous, they should trust their instincts. For instance, odds are good that a reporter for a news outlet that is openly hostile to the idea of global climate change is not going to write an evenhanded article about a researcher's atmospheric chemistry paper. Also, if researchers don't know anything about a news outlet or reporter, they should contact you (the PIO) for information. Researchers don't have to walk into an interview blind.

Once researchers have done their homework, they should write down the two or three key points they'd like to make about their research, limiting each point to one or two fairly short sentences. These "talking points" will help researchers organize their thoughts and will give them a fallback that they can use during an interview (more on this later). Once they've got their key points written down, they can call the reporter back.

Help Them Get It Right

Reporters usually contact a researcher because reporters want help understanding the researcher's work. This means that researchers will have to explain their work to a non-expert. This is when researchers often get nervous. But it's important to note that the reporter does not want to get it wrong. Reporters *hate* getting facts wrong; they want to convey information correctly. And researchers can help them. An easy starting point is for researchers to remember that they shouldn't use words that reporters can't understand. Scientists are often so accustomed to speaking the jargon of their particular disciplines that it is sometimes difficult for them to speak in language that is accessible to non-experts. They should try anyway. Researchers should also keep in mind that some words have different meanings in different contexts. For example, the word “significant” means one thing in a statistical context and another in conversational use. Researchers need to be clear about what they mean.

But to improve a reporter’s odds of getting the story right, researchers need to take steps to make sure reporters understand what they are saying. Here’s one way to do that: after explaining a salient point or a particularly complex issue, a researcher can say, “I want to make sure I’m doing a good job of explaining this. Could you please paraphrase that last part back to me?” English is a delightfully tricky language, and it is difficult to paraphrase something correctly if you don’t understand it. By getting reporters to paraphrase what they have just been told, researchers can often spot a misunderstanding and address it. When I was a reporter, I used to paraphrase big chunks of an interview back to whomever I was interviewing. I caught a lot of mistakes this way, which means those mistakes never made it into my stories. Since most reporters do not initiate this, researchers should take the lead and ask reporters to paraphrase.

Equally important is that by saying, “I want to make sure I’m doing a good job,” a researcher is putting the onus on him- or herself and making the reporter feel magnanimous. This keeps the interview from becoming confrontational, which is probably what would happen if the researcher said, “I want to make sure you’re not going to screw this up.”

How to Answer “Stupid” Questions

Sometimes reporters ask questions that don’t make sense to researchers. This usually happens when reporters don’t fully understand the material. When researchers are asked one of these questions, they can steer things back on course. One option is for researchers to take the time to explain why the reporter’s question isn’t relevant to the work. If researchers have the time (and patience), this is the way to go. But if they don’t, they can take a different approach by answering the question they *wish* the reporter had asked. For example, a researcher could say: “I think the important thing about this research is . . .” and then insert one of his or her talking points. This second technique is called “blocking and bridging." Done well, it's an effective way of conveying the information that is really important.

However, the best strategy might be to take a more straightforward approach. The researcher could simply say, "Look, let me cut to the chase. Here are the things that I think are really important." And then lay out the talking points one by one. Also, there are some questions that researchers shouldn't answer at all: almost all hypothetical questions aren't worth answering. Wild speculation will not help a scientist's reputation.

The Researcher Is in Charge

Lastly, researchers should remember that they are in charge. A researcher can hang up the phone or cut off an interview at any time. The reporter contacted the researcher because, presumably, the researcher has knowledge that the reporter wants. That gives the researcher some control of the process. Ultimately, researchers should try to be patient and help reporters understand their research. Most reporters will appreciate this because they are reasonable and want to do a good job. But occasionally reporters behave inappropriately or irresponsibly. If researchers run into one of these reporters, they should not allow themselves to be bullied into answering questions they don't want to answer. Also, researchers should be wary of speaking "off the record." Unless researchers already have an established relationship with a reporter, they shouldn't say anything to a reporter that they don't want to see in print. Trust can be earned, but it should not be given blindly.

MEDIA TRAINING FOR TV AND RADIO INTERVIEWS

To many researchers, the only thing more terrifying than being interviewed by a reporter for a print article is being interviewed on live TV; live radio is only a little less scary. But if researchers are prepared and don't panic, TV and radio interviews can be very effective science communication tools. It's important to remember that TV is still the most effective way of reaching a large audience. Evening news programs on the major U.S. networks are still watched by millions of viewers—far more people than read even the largest newspapers. But no one is born being comfortable in front of a television camera or a microphone. That takes training and practice.

Identify a Spokesperson

Before issuing a news release or pitching research findings to reporters, it is important to identify one or more researchers who are willing to talk to reporters about the work. The spokesperson is usually the lead author, but that is not always the case.

You should make sure the spokesperson is aware of his or her role ahead of time and that he or she is comfortable talking to reporters and is able to explain the work in language that is accessible to the general public. You should also encourage the researcher to get media training, if you think it's necessary.

Things to Do Before the Interview

As I said in the previous section, researchers should organize their thoughts before an interview and write down two or three key points about the relevant research. It's important to limit each point to one or two short sentences. For TV, they should also be aware that appearance matters. Viewers are less likely to take a researcher seriously if he or she is wearing a T-shirt instead of a dress shirt. If your researchers normally wear T-shirts, sweatshirts, and jeans in the office or around the lab, you might want to urge them to keep a slightly dressier outfit in the closet for unexpected interviews (or visits from funding agency officials). Solid colors are recommended; busy patterns often don't look very good on TV. (Note: Researchers may think that focusing on their appearance is somehow shallow and irrelevant, but it's not. The researchers should want people to be focused on what they have to say, not on what they happen to be wearing.)

Tips for a Taped Interview

Researchers should speak slowly during an interview. When people are nervous (as they're apt to be during an interview), they tend to speak very quickly. This is bad for two reasons: it makes it hard for the news crew to edit the piece because the researcher is running all of his or her words together, and it makes the researcher sound nervous.

Also, urge researchers to avoid jargon if at all possible. Most people have no idea what a rootkit, pulsar, or ganglion is, even though these are basic terms in the fields of computer science, astronomy, and biology, respectively.

It's also important to know that most TV and radio stories will not include the reporter's questions—only the researcher's answers. So train researchers to include the question in their answers. For example, a reporter may ask, "Is this an important advance in our understanding of how global climate change could affect rainfall in southern Africa?" If a researcher answers "Yes," the reporter can't use it, because no one will know what the researcher is talking about. Instead, the researcher should say, "This research is a significant advance in our understanding of how global climate change may affect rainfall in southern Africa."

In addition, researchers should keep their answers short. Most TV and radio stories are fairly brief, so they'll only use short clips of an interview. And by "short," I mean around five to ten seconds—not thirty seconds, and definitely not one minute. The longer the answer, the less likely the reporter is to find a good quote because it's tough to pull a short quote out of a long-winded answer (especially if someone is talking too fast). That means they'll use the quote they can get, which may not be the best quote.

If the researcher made a short list of key points, he or she should use those. This means that whatever quote the reporter uses is more likely to be one you want them to use. Researchers can also tag these quotes by prefacing them with remarks like: "The key point here is . . ." or "It's important to note that . . ." This is especially useful for TV, because sometimes the person who is editing the video is not the person who conducted the interview. The person doing the editing is just looking for a good quote, and those "this is important" remarks act like a giant neon sign saying, "USE THIS QUOTE." Also, if it's a taped interview, researchers should not be afraid to ask if they can answer a question again. Reporters want good quotes, and giving researchers a second chance helps them get one. There's no guarantee that they'll give the interviewee a second try, but it's certainly worth asking.

Finally, researchers should remember that the reporter is not their friend. Most reporters are ethical, patient, and interested in doing a great job, but some are interested in making the story as sensational as possible. When being interviewed, researchers should always assume that the camera and microphone are on. That little lapel mic that they put on the researcher's collar? It keeps recording until the news crew turns it off, regardless of where the camera is pointing. Remind researchers not to say or do anything they wouldn't say or do on the air—because it just might end up on the air.

Tips for a Live Interview

The good news about a live interview is that researchers usually have more than five seconds to answer any questions a reporter might ask. Researchers also don't have to remember to incorporate the reporter's question into their answer. The bad news is that there are no do-overs. As a PIO, your job is to make sure that researchers are prepared. A live interview flows a lot like a conversation. The reporter will ask questions, and the researcher will answer them. But, like a conversation, a live interview also gives your researcher some flexibility. If the reporter asks a question that's off-topic, the researcher can change the flow of the interview. A researcher could answer, "Our study didn't look at that issue, but we did find that [insert one of your prepared key points here]." If the reporter is any good, this can lead to an interesting new direction for the interview and can help steer it back onto solid ground. As I said in the previous section, researchers shouldn't answer hypothetical questions—speculating is dangerous.

The key message in media training for researchers is: Don't panic. The interview is about a subject the researcher knows a lot about; that's why the researcher is being interviewed. The researcher almost certainly knows more about the subject than the person conducting the interview. So while researchers shouldn't act smug, they should be confident. After all, they are the experts.

More Practical Tips

Two quick, but useful, tips:

- When being interviewed on TV, researchers should look at the person who is asking the questions, not into the TV camera.
- If standing during the interview, researchers should not shift their weight from foot to foot; it will look like they are standing on a boat.

A PIO'S ROLE DURING INTERVIEWS

So far, this chapter has focused on what a PIO should do to prepare researchers for news interviews. But how should a PIO prepare for an interview? And what should a PIO be doing during the interview? It depends on the researcher, the reporter, and the institution you work for.

Questions in Advance

Many institutions require PIOs to ask reporters for questions in advance. Reporters don't like this, but it can be useful. Asking for questions ahead of time can give you a good idea of what the reporter is interested in talking about in order to ensure that you've set up an interview with the correct researcher. For example, if a reporter has contacted you in regard to a paper about hurricane prediction and "big data" computer science, you'll want to know whether the reporter wants to talk to a computer scientist, a meteorologist, or both.

However, if you are requesting questions in advance so you can try to control the direction of the interview, you may be less successful. Good reporters will have done some homework before the interview but will still go into an interview with an open mind. They'll come up with new questions based on the answers they get from your researchers. Pulitzer Prize–winning reporter Deborah Blum once told me, "I'll do research and I'll write down questions, but I also think a good interview is akin to good conversation, and if you're too rigid in your prep work, too obsessive about your written questions, you lose those moments where the story may open up into something more."

If you do request a list of questions in advance, don't expect reporters to send anything more than a list of short, general questions—and don't be alarmed if a reporter goes off-script during the interview. As long as the questions are reasonable and the reporter doesn't belabor the issue if a researcher declines to answer, there's no reason for a PIO to intervene.

PIOs Sitting in on Interviews

Many institutions require PIOs to be present whenever a researcher is being interviewed. This gives the institution some oversight over the process, can help put researchers at ease, and allows PIOs to intervene if a reporter badgers a researcher. That said, I very rarely sit in on interviews. The researchers I work with are all capable of handling themselves, and they know more about the interview topic than I do. However, I am a PIO for a university with an active research program, as opposed to an industry organization, a federal agency, or a private company. Those types of institutions are, generally speaking, much more likely to want PIOs to sit in on interviews, for a variety of reasons.

There are times when I do sit in on interviews. This is usually because the researcher has asked me to. My role in these situations is to provide moral support for the researcher. I don't think there's anything wrong with my being there, as long as I don't interject and cut off the researcher or the reporter. Sometimes my presence is actually beneficial to the reporter. If questions come up that the researcher can't answer, I can sometimes connect the reporter to other researchers who might be able to help. I can also help reporters find (and access) university facilities that can be used to create photographs or video footage to accompany the story. But PIOs should not answer questions unless the questions are directed to them. You are there to observe the interview, not to be part of it.

Lastly, PIOs should inform reporters about the conditions of the interview ahead of time. If you'll need advance questions or you plan to sit in on the interview, give them plenty of notice. It's disconcerting and impolite to blindside a reporter with requirements at the last minute.

KEY POINTS

- Have research teams identify a spokesperson even before being contacted about an interview (if possible).
- Before an interview, have researchers familiarize themselves with the reporter and the news outlet.
- Have researchers write down the key points they want to make—and keep them short.
- Coach researchers on how to help a reporter understand their work by saying something like: "I want to make sure I'm doing a good job of explaining this," and asking the reporter to paraphrase what they have said.

- Have researchers highlight key points in the interview: "The important thing here is . . ."
- Remember that the researcher has control in an interview.
- For TV interviews, researchers need to know that appearance is important.
- Remind researchers to include the question in their answer.
- Tell researchers that they shouldn't answer hypothetical questions.
- It's okay to ask a reporter for questions in advance, but be reasonable.
- It's okay to sit in on an interview, but do not become part of the interview.
- Make sure reporters know about any interview requirements in advance.

6

TELLING THE STORY YOURSELF

SOCIAL MEDIA AND BLOGS

Some people view social media as a waste of time. Some view social media as an enormously valuable suite of communication tools. Both views are correct—it really depends on how you use social media. This chapter will focus on how to use social media (including blogs) effectively, the pros and cons of various social media platforms, and useful tips for how to build and sustain a following online.

Technological innovation and user trends make the social media landscape extremely dynamic. This poses a challenge, because I'm writing this book months or years before you will read it. To address this issue, I'm going to focus on broad themes and long-term trends, rather than trying to address minute details that will almost certainly have changed by the time this book comes out. My goal here is to help you think about social media in a critical way so you can use these tools effectively. Once you're thinking critically about social media, you'll be able to make good decisions about how to use these platforms and respond to changes as they come up.

WHY SOCIAL MEDIA AREN'T (NECESSARILY) A WASTE OF TIME

Scientists, PIOs, and research institutions can all benefit from active engagement with social media. But not everyone believes this to be true; many people think that Twitter, Facebook, and other social media platforms are used solely to tell the world where you ate lunch or what cute things your cat has done recently. This is because there *are* people who use social media solely to tell the world these things. But—and this is important—no one can make you post trivial things on social media. And—this is even more important—no one can make you follow people who post things you don't care about on social media.

Social media is a catch-all term used to refer to a variety of communication platforms. Those platforms do not control content; users control content. E-mail can be used to send someone a long list of knock-knock jokes. But most people have accepted the fact that e-mail has practical utility. Researchers can use e-mail to share information with colleagues and peers about grant opportunities, new research findings, or job openings. The same holds true for social media. In short, social media are tools that can be effective components of your communication efforts.

Social Media Are Not E-mail

E-mail is an effective communication tool, but it only works if you know precisely whom you are trying to reach. Social media, on the other hand, can be great tools for engaging a larger community of people, most of whom you don't know. If you become part of an online community that is relevant to your work, you can tap into the experience and expertise of a lot of people in your field that you don't already know. Trying to figure out which science reporters are interested in cancer research? Ask the online science-writing community—it is both large and welcoming. Networking doesn't just happen at conferences anymore; it happens online every day.

Tracking Trends

You can also use social media to track trends in popular topics of discussion. A research finding that might seem dull at first can take on new life if it relates to a subject that is in the public eye. Similarly, you can use social media to find out what reporters are interested in. Twitter, in particular, is useful for pinpointing reporters who have expressed an interest in a specific subject. If you were preparing to pitch a story on forensic anthropology, you could search for related terms on Twitter. Have any reporters been tweeting about the subject recently? If so, you can look them up online and see if forensic anthropology is a subject they write about. If they do, you should add that person to your list of reporters to pitch forensic stories to.

PICKING A PLATFORM

There are dozens of social media platforms on the market, each with its own set of advantages and disadvantages. One of the first things you need to do when developing a social media strategy is to determine which platforms will help meet your institution's communication goals. To that end, here's an overview of the pros and cons of some of the more popular social media platforms. I'm starting with Facebook, the most omnipresent social media platform, and closing with Twitter—which I think is likely the most important platform for PIOs at research institutions.

Facebook

Even most people who have no interest in social media are familiar with Facebook. It allows institutions to create "pages" and share Web links, text, images, and video, giving users a great deal of flexibility in terms of the type of content they can share. Facebook reaches an audience of more than 1 billion users. It's hard to pin down the precise number of "active" users who engage with the platform on a regular basis, but in 2013 Facebook was reported to have more than 750 million active users globally and around 200 million monthly active users in the United States. Those numbers are always changing, but they give you an idea of Facebook's overall reach. That flexibility and potential audience make Facebook enormously attractive to institutions. However, there are some things to bear in mind when creating or managing a Facebook page for your institution.

If you create or manage a Facebook fan page for your institution, you are posting images, videos, and links related to your employer. The goal is to disseminate information to fans, or people who "like" your page, in order to foster a feeling of connectedness between your institution and its stakeholders. If you work at a research institution, it is likely that many of those fans will be researchers, science reporters, or other key stakeholders in the research community. Hopefully, your fans will then share your posts with their connections, broadening the reach of your communication efforts. But it doesn't always work that way.

When you post something in your capacity as a page administrator, that post will appear on your institution's page, but the post will show up in the news feeds of only a fraction of your followers. Facebook uses an algorithm to determine which followers' news feeds will see the post. The algorithm is dynamic, so as more people interact with the post ("liking" or commenting on it) the post will appear in the feeds of more users. This can lead to a “feast or famine” situation: posts that are not quickly noticed and liked by your followers go ignored, but posts that are noticed benefit from a cascade effect because garnering “likes” expands the reach of the post, drawing more interaction, which expands the reach further, and so on. In addition, Facebook changes its algorithm fairly often. So a strategy that helps you reach your audience today may not work very well next week. If you want the post to appear in all of your fans’ news feeds, you have to pay a fee to “promote” the post. If your fan page has a large number of followers, this can quickly become expensive.

You should also be aware that Facebook users often don’t want to leave Facebook to view information you point out to them, so posting links to other sites may not be an effective strategy. Instead, you might have more success using infographics, photographs, or videos that allow users to view what you want them to see without leaving the platform. Depending on the resources you have available, that may not be easy.

In summary, Facebook is a useful tool with a great deal of potential. However, its algorithm-driven system for sharing posts means it is an inconsistent way of reaching your followers, unless you can pay to promote your posts—and your posts may not make much of an impact if they ask users to leave Facebook.

Google+

Google+ is similar to Facebook in that it allows you to create a page for your institution and share a wide variety of content: text, links, video, and images. Although Google+ does reach fewer people than Facebook, it is still one of the largest social media platforms on the planet. Estimates vary, but Google+ was reported to have more than 350 million active users in 2013.

Google+ is also appealing to institutions because everything you post shows up in your followers’ news feeds. Getting “+1s” (which are equivalent to “likes”) can boost a post’s profile by returning it to the top of a news feed, which also makes it easier for unpopular posts to get buried. Yet your followers will at least have the opportunity to see everything you post.

Google+ also has two additional advantages: search engine optimization (SEO) and Google+ Hangouts. Because Google+ is owned and operated by Google Inc., posting content on Google+ can improve your institution's SEO on the Google search engine. SEO is the process of modifying your content or content presentation in order to make it more visible in search engine results. Here's how Google+ comes into play in SEO: if you post a link to your institution's website on Google+, Google is able to track every time users share, +1, or comment on that link. For example, if you post a link to a blog post you wrote about black widow spiders and that post garners a significant amount of attention on Google+, your link's profile will get a boost in Google's search engine results.

Google+ Hangouts allow multiple people to interact in real time. This has a great deal of potential for PIOs. You could set up an online interview between reporters and a researcher, allowing the researcher to give a presentation with images and live audio. You could also hold a live video press conference or arrange for a team of researchers to field questions from the public in a live online forum.

Google+ is a useful tool with a great deal of potential, and it will likely only become more useful if the number of Google+ users continues to grow.

YouTube

If your institution creates video content, a YouTube account is a virtual necessity. YouTube allows you to create a channel that features your institution's videos and is probably the best platform for sharing videos with the public. In late 2013, YouTube reported receiving more than 1 billion unique visitors each month. Those visitors watched a total of more than 6 billion hours of video, and 30 percent of that traffic came from within the United States. In short, YouTube gives you the potential to reach an enormous number of people.

The best way to attract an audience for your videos is to produce good videos. I addressed this in the earlier section on multimedia, but here's a brief recap: keep your videos short and visually compelling. To increase the odds that people will find your videos, you should also ensure that your videos are listed under a relevant category and include accurate (but interesting) titles and clear descriptions. You should also make good use of "tags"—keywords that describe the video's content—which YouTube uses to help connect users and content. For example, if you work at the National Institutes of Health's National Cancer Institute and you post a video to promote breast cancer screening, you might want to list it under the category "health" or "science and technology"; title it "Breast Cancer Screening: Early Detection Saves Lives"; and include a concise description of the video. You would also want to include these tags: NIH, National Cancer Institute, NCI, cancer, breast cancer, screening, prevention, and awareness. You may add more tags, depending on the content of the video.

Also, if your institution receives any federal funding, remember that you need to comply with the Americans with Disabilities Act regarding accessibility of your published materials. If you post videos that include narration or dialogue, you will need to include an accurate transcript with the video when you post it online. YouTube offers easy step-by-step instructions for uploading a transcript. Other than typing the transcript, this step is neither difficult nor time-consuming.

Even if you don't expect users to run across your videos while browsing YouTube, there are several reasons you should create an account and post your videos there. First, YouTube has a somewhat symbiotic relationship with other social media platforms. Facebook and Google+ users like to click on video links, which improves your engagement on those social media platforms. When they click on those links, you are driving them to your YouTube channel. By using these social media tools in tandem, you are giving your audience compelling content (which they want) while also driving users from one of your institution's sites to another (which you also want because it means you have their attention). Second, putting videos on YouTube allows you to either embed those videos or include hyperlinks to them in your news releases and blog posts. This can make your releases and blog posts more compelling (assuming the videos are good), and it gives you more face time with your audience. Again, you are holding the attention of your audience. Finally, YouTube is a convenient way to share videos with reporters. It's much easier to include a link in an e-mail than to send large video files.

Twitter

Twitter allows users to post messages of 140 characters or fewer, including links to other sites, videos, or images. Twitter reaches fewer people than Facebook, Google+, or YouTube; according to Twitter's IPO filing, in 2013 it was reaching approximately 49 million monthly active users in the United States. Despite its limitations, in my opinion Twitter is the most useful social media platform for science PIOs. When you "tweet," or post a message on Twitter, all of your followers have the opportunity to see your message. Twitter is also easy to search, allowing users to find your content. But there are really four important things that science PIOs should consider about using Twitter: Tweets can go viral; research communities use it; reporters use it; and you can use it to establish yourself as a reliable source.

Twitter is capable of spreading information with amazing speed. A tweet with a catchy line, a captivating image, or a link to a compelling story can be shared (or "retweeted") quickly and easily. If even one prominent Twitter user sees your tweet and passes it on, it can create a cascade effect, with dozens of users sharing your tweet with all of their followers, who share it with their followers, and so on.

More importantly, Twitter can help you reach the *right* users. That's because Twitter is home to a large number of informal research communities. These are unofficial, often overlapping, networks of people with a shared interest in a specific subject, whether that subject is entomology, public health, or science communication. If you want to disseminate information to people who are interested in a specific research area, you can become part of one of these communities. This isn't as simple as signing up for a service. It involves interacting with other users on a regular basis, engaging with other members of the relevant community, and sharing interesting and relevant information. But once you've become part of the community, you can spread information very quickly.

For example, if you are trying to raise awareness among entomology researchers about a new research initiative, you could tweet a link to blog post about the project. If your message is retweeted by the Entomological Society of America or by a prominent entomology blogger, you could reach hundreds or thousands of entomologists very quickly. And other researchers won't be the only people seeing those tweets. Reporters often follow prominent members of these research communities. If a science reporter has an interest in covering stories about bugs, they probably follow prominent

Other Social Media Platforms

In the interest of brevity, I've focused on the most high-profile social media platforms. However, there are a number of other social media platforms that you may be interested in exploring. Pinterest, Tumblr, and Flickr are all excellent sites for engaging broad audiences, particularly if your institution creates a steady stream of compelling, research-related images. They may not be particularly effective tools for reaching science reporters, but some news outlets include features that focus specifically on images, and these sites may also be useful for reaching other audiences. LinkedIn is a useful site for personal networking but is of limited value in terms of raising your institution's research profile or reaching reporters.

I encourage you to explore new social media tools if you have the time and interest. You may find something that works extremely well for your institution. But don't be afraid to drop a new platform if it's not working for you. I've experimented with more than a dozen social media platforms, but I use only a handful on a regular basis.

entomology bloggers and researchers on Twitter. If those entomologists are retweeting your posts, relevant reporters will see them too. Reporters will take good story ideas wherever they find them. In addition, many reporters turn to Twitter when doing research on stories. Reporters may search Twitter to see who is talking about a specific subject or to find potential sources for a story. If a reporter is plugged in to any of the informal research communities on Twitter, they may reach out to those communities to find experts on related subjects.

Finally, Twitter is the only social media platform where science PIOs may want to work on both their institution's account and their personal accounts in a professional capacity. An institution's Twitter account is useful for promoting research findings, news releases, and blog posts. Your institution's account can also be used to highlight news stories featuring the institution's researchers. But PIOs may also want to consider using their personal Twitter accounts to cultivate relationships in online research communities or with reporters. You don't want to clog your institution's Twitter feed with links or conversations that are not specifically related to the institution, but those are exactly the things you have to do in order to engage with researchers and reporters. If you become an active member of these informal Twitter communities, researchers and reporters are more likely to retweet your posts, respond to your questions, or otherwise interact with you online. Active participation in informal Twitter communities can also raise your profile in the science reporting community. Reporters who won't open an e-mail pitch from a PIO they've never heard of may be more likely to read what you send them if they're familiar with your name from Twitter.

FIRST STEPS: KNOW YOUR TOOLS AND BUILD A FOLLOWING

If you're using a social media account, your first step should be to familiarize yourself with how the relevant social media platform actually works. I'm not talking about effective communication techniques; I'm talking about the basic ins and outs of the site. You should know that when you paste a link into Facebook, you can alter the headline by clicking on it. Similarly, you should know that other users cannot send you a direct message on Twitter unless you follow them. Mastering the basics is important because you have to be aware of what you *can* do on a social media site before you can determine what you *should* do on that site.

Once you learn how to use a platform, your next step is to figure out how to reach the audiences that are important to your institution.

Building a Following for Your Institution

Just because you set up a social media account doesn't mean people will automatically know about it. You'll need to take the time to cultivate a following. To do that, you must remember that social media are *social*. So if you want to build a following, you need to engage with people. You have to give other users a reason to follow you—and that means you can't simply go online and post a series of links that are all about your institution. You should also share posts by other users that are relevant to both your institution and the research community at large. Comment on the posts of other users, when appropriate. Be responsive if other users ask you questions.

If you work at a large research institution, you will have a head start over other users and should be able to garner a certain number of followers without much effort. For instance, some reporters and stakeholders will follow government agencies, and alumni will often follow their universities. But that audience is limited. And even they won't pay attention unless you're sharing engaging content. That is the most important thing you can do to cultivate a following: make sure you're posting links, video, or images that people will be interested in. If you engage with others and share compelling content, followers will find you.

Once you have followers, you can't rest on your laurels. It is easier to lose followers than to gain them. You have to continue to be engaged and provide a steady stream of good content to keep your old followers and to attract new ones.

Building Your Following on Twitter

Building a following for a personal Twitter feed is different from building a following for an institutional Twitter feed because you have more flexibility in what you post and how you interact with other users. I think it's useful for science PIOs to have their own Twitter accounts, so I want to talk about things you can do to develop your own following.

First, you should tweet about things that you are actually interested in, but avoid sharing overly personal information on your Twitter feed if you plan to use the account for professional networking. It's okay to mention your interests—you don't want to come across as an impersonal robot—but you probably don't want to talk about your sex life, your in-laws, or substance abuse. And you don't want your personal tweets to significantly outnumber any tweets that relate to your professional life. I'll occasionally tweet on my personal Twitter account about my fondness for jazz or science fiction, but the bulk of my tweets are about research and science communication. Since I'm genuinely intrigued by those topics, my Twitter posts are sincere. If you fake an interest in something, it's usually obvious. So don't pretend to take an interest in something you don't care about in the hope that it will help you professionally.

You should also be upbeat when writing for Twitter. A 2013 study found that tweets containing positive sentiments (e.g., "This study is really interesting") are positively associated with gaining followers.[1] By the same token, tweets with negative sentiments (e.g., "This study is a waste of time") are negatively associated with gaining followers.

Remember to provide useful information to your followers. Sharing good content—links to news stories, job announcements, or new studies—works to your advantage. Finding good content is a big reason why many in the science community join Twitter in the first place. But avoid being a so-called "meformer," who tweets a lot about him- or herself. The 2013 study found that "informational content attracts followers with an effect that is roughly *thirty times higher* than the effect of 'meformer' content, which deters growth" (emphasis theirs).

Further, the researchers found that it's okay to use big words—as long as you spell them correctly. If you sound smart, you are more likely to get followers. Or, as the study authors put it, "Twitter users apparently seek out well-written content over poorly written content when deciding whether to follow another user." As a writer, you should be overjoyed to hear this. There's also good news for Twitter users who really only want to talk about one subject, whether it's astrophysics or oncology: the study found that focusing on specific topics also helps to attract new followers.

You should also be sure to fill out the profile section of your Twitter account. The 2013 study found that people are more likely to follow your Twitter feed if you have taken the time to explain who you are on your Twitter profile—especially if your profile includes a URL. The researchers said that completing the user profile may help persuade other users of "one's authenticity and trustworthiness, making them more likely to become followers." Lastly, use a photo of yourself in your Twitter profile. A 2012 study found that Twitter users think that Twitter feeds featuring photographs as the profile

1. C. J. Hutto et al., "A Longitudinal Study of Follow Predictors on Twitter" (paper presented at the Computer Human Interactions conference, Paris, France, April 27–May 2, 2013).

pictures are more credible than those that use cartoons or other images.[2] Twitter feeds featuring the default Twitter "egg" image as the profile picture were considered the least credible.

MANAGING YOUR SOCIAL MEDIA ACCOUNT

Science PIOs don't use social media accounts to amass followers. We use social media accounts to achieve communication goals. That means you'll need to manage your social media accounts effectively.

What to Post

What you post on your social media accounts depends on whom you are trying to reach and what you are trying to accomplish. Specifically, you want to focus the social media content you share on the needs and interests of your target audience. If you work at a university trying to attract high-quality researchers in a specific field, you may want to focus on sharing content specific to that field. This could include news on research findings in the relevant discipline, announcements of grants that your university has been awarded, and blog posts about new research equipment and facilities. You may also want to include news about the quality of life in the region where your institution is located. Similarly, if you work at a public research institution and want to raise your profile in a positive way with political figures who control your budget, you'll want to focus on content that appeals to politicians. This might include the practical applications of your research, such as improving public health or boosting energy efficiency. You might also want to highlight stories about the economic impact of your institution and the institution's ties to the local community.

In short, you need to engage in some critical thinking. Ensure that your social media efforts make sense in the context of your larger communication goals and that they are consistent with your institution's "brand," or how it sees itself. Know what your institution wants to be, and coordinate your social media activity with your

2. Meredith Ringel Morris et al., "Tweeting Is Believing?: Understanding Microblog Credibility Perceptions" (paper presented at the 2012 ACM Conference on Computer Supported Cooperative Work, Seattle, WA, February 11–15, 2012).

other outreach efforts so that all of those communication elements work together to support your institution's brand.

You also need to think about what forms of content you'll be sharing. Earlier I noted some of the strengths and weaknesses of various social media platforms, but it's worth revisiting them briefly. Facebook, Google+, and Twitter are all most effective when you're driving users to meaningful content. That content can consist of news releases, blog posts, or news stories from outside news outlets. But for Facebook and Google+, it's important for your posts to have a visual element. Links with associated photographs or other engaging images are much more likely to garner "clicks" from users on those platforms. Links with no associated images aren't likely to get much attention nor are text-only posts. Videos can be popular posts, but they don't tend to be as popular as photographs or other still images.

For Twitter, it's a good idea to use hashtags judiciously and sparingly. A hashtag is a search term entered into a tweet by typing the # sign followed by a word (e.g., #SciencePIO). This allows users to search for all tweets containing that hashtag simply by clicking on it. These are widely used by people tweeting content related to a specific topic or event. For example, the hashtag #sciwri13 indicates that a tweet is about the 2013 National Association of Science Writers conference. Identifying and using hashtags that are popular with a target audience can be an effective way of reaching users who do not follow your Twitter feed. If you're promoting research findings that are being presented at a conference, you could include the conference's hashtag in your tweets about that research. That makes it more likely that other attendees of the conference will see the tweet—and could even improve turnout for your researcher's presentation. However, hashtags are often overused. If you put a hashtag in front of every noun in your tweet, it's annoying and doesn't highlight specifically what your tweet is about. In short, you are actively hurting your social media efforts. You should use hashtags only to highlight one or two keywords to reach a specific audience.

Lastly, don't forget that social media should be social. If you want to engage with other users online, you need to be interactive, but you can't be interactive if you're only posting photos, videos, and links. It's a good idea to share interesting content posted by other users, to respond to queries from other users, and to post questions of your own. If you work at a university that is trying to engage with an audience of students and alumni, ask followers to share something about their favorite class or professor. If you're trying to reach an audience of astronomers, ask them to tell you what drew them to the stars in the first place or to make an argument for why their favorite celestial body is the best one in the sky. Don't be afraid to do something that's just for fun. This sort of dialogue may not be directly in line with your institution's communication goals, but it does help to build a sense of community, and that lays the foundation for your more strategic communication efforts.

Timing Your Posts

Showing up in your followers' news feeds is no guarantee that they'll see what you post. It simply means they have the *opportunity* to see it. If you post something at 5 p.m. on a Friday evening, your followers probably won't pay attention. You can usually determine the best times to share posts using trial and error. Do early morning posts do well? If so, keep posting things in the early morning. If not, post them at another time and see if that works better. Some platforms, including Facebook (at least for the moment), have analytics that tell you specifically when most of your followers are active on the platform. I have found that the audiences I want to reach tend to be most active early in the work day—just after people have gotten to the office—and around lunchtime. However, your audiences may be on different schedules. Figure out what works for you and stick to it. Also, don't be afraid to post something more than once, particularly on Twitter. You don't want to share the same item over and over again; that would quickly become annoying. But it's perfectly reasonable to share a post early in the morning and then share it again late in the afternoon. Odds are good that a different audience will be online and you'll be able to reach a new group of people.

You also want to space out your posts. Posting three or four things in rapid succession does two things, both of them bad. First, it floods your followers' feeds, which is annoying and can make people unfollow you. Second, people are unlikely to click on *all* of the things you posted, so you're wasting your effort. If you have an enormous amount of time-sensitive content that you want to post, wait at least an hour between posts. If possible, you may want to separate them even more. One reason to space out your posts is because you want to budget your content to be sure that you're able to share interesting or important news on a fairly regular basis. You don't want to go days or weeks without offering valuable content to your followers.

Lastly, I recommend that you not use tools to automatically schedule when your posts go live. Scheduling a post can be useful for posting content over the weekend or when you're out of the office, but it also creates the potential for unforeseen consequences. For example, a research institution might schedule a seemingly innocuous tweet to go up at 10 a.m. on a Saturday morning. The tweet could say, "It's a great day to be in the Gulf of Mexico" and link to a story about marine biology research. However, if a cruise ship sinks in the Gulf of Mexico at 8 a.m. that Saturday, drowning dozens of people, the tweet would seem tone-deaf and out of touch, and I can almost guarantee that it would lead to a story about "poor social media practices" featuring your institution as the poster child for doing it wrong. If you do choose to schedule posts, make sure that multiple people have access to the social media account and can prevent untimely posts from going live or delete posts if they've already gone up.

Pitching to Reporters on Social Media

I recommend that science PIOs pitch to reporters via e-mail, but that's not always possible. In many cases, it's hard to find a reporter's e-mail address, and you want to reach the reporter in a timely way. In those instances, pitching to a reporter via social media can be effective. Just as with e-mail pitches, it's important to know what a reporter covers. Don't pitch them stories that are outside their beat. (For more on this, refer to chapter 3.)

If you're going to pitch to a reporter via social media, do it on Twitter. This is a public platform, and it's easy to find and reach out to reporters. Don't try to "friend" reporters on Facebook just so you can pitch them stories. I'm friends with reporters on Facebook, but I don't pitch to them there. I am Facebook friends with those reporters because I know and like them, not because I want to take advantage of them. To pitch to a reporter via Twitter, you can go to their Twitter account page and type in the box that says "Tweet to [Reporter's Name]." It's that simple. But before you pitch to a reporter on Twitter, make sure you follow their Twitter feed. This is important because a reporter can't send you a direct message unless you follow them. Even if they want you to send them additional information, they probably don't want to post their e-mail address or phone number on their public Twitter feed (or else they'd have included it in their profile).

Don't be a pest. If you've sent a reporter a tweet but haven't heard back, assume he or she is not interested. Badgering reporters will not get you results—or, at least, it won't get you the results you want. And it may even cause them to block your account completely. Don't send the same pitch to three or four different reporters. Twitter is a public forum, and everyone will be able to see what you've done. If it looks like you're trying to sell the same story to everyone you can think of, a reporter is less likely to follow up with you.

Other Content-Sharing Sites

In addition to social media platforms, there are a number of other content-sharing sites that can drive traffic to your blog or institutional site. The most famous, and infamous, of these content-sharing sites is reddit. Once you create an account on reddit, you can post links to almost anything. You can also classify your posts according to various subreddits, which are category-specific pages within reddit. Other reddit users can "upvote" or "downvote" your submissions. The more votes your post gets, the higher its visibility on the site.

If you're sharing an interesting post, have given it a good headline, and have posted it on the appropriate subreddit, you can reach an extremely large number of people. I've seen instances where reddit posts have driven thousands of visitors to sites in only a few hours. However, you need to be familiar with reddit's protocols. If the bulk of the links you share are links to your own site (or your institution's site), you'll be banned from the site for spamming. You need to post links to lots of different types of content in order to maintain your reddit account.

Similarly, be sure to comply with the submission guidelines of each subreddit; they're prominently featured on the right-hand side of each subreddit's landing page. If you post links in violation of

Social Media Tools and Critical Thinking

The goal of this book is to help you work effectively as a science PIO. I provide some practical advice, but more important than giving advice is getting you to think critically about what you're doing and why you're doing it. That is true for every aspect of the job, but it is particularly relevant for social media because the social media landscape changes very quickly.

In short, the tools at your disposal are constantly changing or being taken away altogether, even as new tools are being added. My goal for this chapter is not to make you dependent on the tools. Instead, I'm hoping this chapter can help you focus on your goals and find the tools you need to meet those goals.

those guidelines, your posts will be deleted. For example, the science subreddit requires that all posts be direct links to peer-reviewed research or a summary of peer-reviewed research. If you post a link to a story about new grant guidelines for the National Science Foundation, your post will be deleted. It's related to science, but it doesn't meet the subreddit's guidelines. Also, while reddit can be a great place to promote research news, it is also home to some unsavory behavior. Some subreddits and reddit users share blatantly offensive material, including sexist, racist, violent, and pornographic content. As with the Internet in general, you should tread carefully.

Other content-sharing sites include Slashdot and Hacker News, which are great platforms for sharing technology news, as well as StumbleUpon, Digg, and Fark. All of these platforms give you the opportunity to reach a lot of people, but you need to familiarize yourself with each site's rules and the audience that each site can reach. If you do your homework and follow the rules, content-sharing sites can be an effective way of amplifying your reach.

BLOGS

Blogs can be a valuable resource for a science PIO, allowing you to dig into the nuance, context, and detail of a subject in a way that you can't with news releases. It also gives you the opportunity to explore facets of news stories that have been ignored in other outlets, to discuss papers that may have been overlooked, or simply to share anecdotes that highlight the odd, fun, or funny aspects of scientific research. But blogs can also devour an enormous amount of time and resources. This section will help PIOs determine how they might be able to use a blog and whether they have sufficient time and resources to make a blog a worthwhile investment.

Things You Can Do with a Blog

I wrote in chapter 2 about the flexibility that blogs give you in terms of narrative style. But blogs also allow you to write about things that you wouldn't write about in a news release. Here's a list of things that a blog allows you to do:

- Write a post about interesting research findings that are too old to merit a news release.
- Highlight interesting grants that aren't large enough to merit a news release.
- Write features that offer a new perspective on external news stories.
- Write features that offer in-depth explanations of how research works—information that news releases often gloss over.
- Run interviews with researchers.
- Run guest posts written by researchers.
- Write posts about new facilities or equipment and their impact on your institution's research.
- Write extremely short posts that focus on compelling research images or video.
- Write posts that tap into your researchers' expertise to explain the science behind everyday questions, such as how bees make honey or how 3-D printing works.
- Focus on new applications of previous research.

All of those things can make for good stories. They can also help you reach target audiences and advance your communication goals. You can drive traffic to those types of posts through social media or content-sharing sites. But none of those things would fit neatly into the news release formula.

How to Use Your Blog

One way to use your blog is to garner media attention. You could write posts that highlight a researcher's expertise in a specific field by running an interview with the researcher, writing about a researcher's insights into an external news story, or getting the researcher to write a guest post about his or her work. Whichever approach you take, you can use the relevant post to reach out to reporters. For example, if a researcher is an expert on flu vaccines, you could contact health reporters in the early fall and let them know that the researcher is available to serve as a source if they'll be writing a piece on flu season. Your pitch should include the researcher's contact information and a link to the blog post. By providing meaningful, in-depth content, the reporter can make an informed decision about whether to use your researcher as a source. But make sure your post is actually informative. If it's a puff piece that says the researcher is great but doesn't include any details, there's no reason for a reporter to take you (or the researcher) seriously. Ideally, you could also use the post to reach other audiences via social media.

You can also use your blog posts to pitch stories to reporters. A news release is generally aimed at a large number of reporters, but a blog post can be about subject that you know most reporters don't write about. If you know of any reporters who do cover that subject, either because they write for a targeted audience or because they have a personal interest in it, pitch to them exactly the same way you would with a news release (see chapter 3). The only difference is that you'll include a link to a blog post instead of a link to an online news release. I've used this approach to pitch stories that ended up running in venues ranging from the *Wall Street Journal* and the *Los Angeles Times* to *Popular Science* and *Scientific American*.

You can also use your blog to tell stories that will appeal to specific audiences other than reporters. If you're trying to attract new researchers to your institution, you can write a piece about state-of-the-art research facilities. If you want to entice high-quality students to a university, you can run a series of posts about students who have made significant research findings or about opportunities for students to engage in innovative, hands-on research.

You should also make sure that blog curators know your blog exists. The National Science Foundation has a news site called Science360 that is updated daily with news from research institutions, media outlets, and blogs. You should contact the NSF to make sure they're aware of your blog so that they can feature any of your blog posts that they deem important. Science360 does not receive a large amount of traffic, but the people who do read it are your target audiences—researchers, funding agency officials, and administrators from research institutions.

Lastly, you should be aware that other entities in your institution may be able to use your blog content. Posts on the research blog at North Carolina State University are often also featured on the university homepage, college and departmental pages, or in other university publications. If you're creating valuable content, people will want to use it.

Time and Resources

I've explained how a blog can be a useful tool for science PIOs, but that doesn't mean every institution should start a blog. You need to know whether you'll have the time and resources to make a blog successful. Maintaining a blog means providing a steady stream of content. You don't have to update the blog daily, but you should be able to post a new story, video, or image at least once a week. If you can't commit to doing that, you may not want to start a blog in the first place. The question you have to answer before launching a blog is this: Do you have your boss's support to make time in your work schedule to find, write, and post content to a blog on a regular basis? It helps if you work for an institution that has multiple PIOs who can all contribute to the blog. This allows you to divide the workload. But ultimately you need to have the backing of your supervisor. If he or she does not think it's a valuable use of your time, you won't be given the hours you need to do a good job.

Setting Up a Blog

If you do decide to create a blog, ask for support from your institution's Web or IT staff. If possible, you'll want to incorporate the blog into your institution's website. There is a wide variety of blogging platforms you can choose from, with WordPress and Blogger being two of the most popular. However, you'll want to make sure that the blogging software you use allows you to integrate the blog into the rest of your institution's site. The blog should be clearly connected to your institution, and working with professional Web designers can help to ensure that the software and design options you choose allow you to create a blog whose visual identity is consistent with the visual identity of your institution.

You'll also want to create categories and tags that are consistent with the work that your institution does. For example, if your institution engages in both medical research and clinical medicine, you may want to have a category for each. You could then create tags that focus on specific medical disciplines: oncology, pathology, and so on. This will help you organize your posts in a meaningful way and allow users to efficiently seek out the information they're looking for.

KEY POINTS

- Social media can be useful tools, if you use them effectively.
- Familiarize yourself with various social media platforms, and decide which ones can help you meet your communication goals.
- You almost certainly want an institutional Twitter account, as well as a personal Twitter account you can use for networking.
- Remember that social media are *social*, which means engaging with other users.
- Remember whom your target audience is, and share content that is relevant to them.
- Make your social media posts positive and intelligent.
- Timing is everything: don't post multiple things at once, but don't go days between posts either.
- Think carefully before scheduling posts. A poorly timed post can turn into a public relations nightmare.
- Don't be afraid to pitch to reporters on Twitter, but be smart about it.
- Make use of content-sharing sites, such as reddit, but be sure you follow each site's user guidelines.
- Blogs allow you to do things that you can't do with news releases.
- Don't start a blog if you don't have the institutional support, time, and resources you need to make it work.
- If you do decide to start a blog, work with your institution's Web professionals.

7

MEASURING YOUR STORY'S SUCCESS

METRICS

You're writing news releases and pitching stories to reporters. You've started a blog and are using social media to push your posts out, and you know that people are reading them. But you need to have some way of measuring whether you're making progress toward your communication goals. It's important to be able to determine whether any of your communication efforts are successful; and if they're not working, you need to rethink what you're doing. You need metrics—tools you can use to measure your progress. This chapter will discuss various metrics you can use to determine the success of your media outreach and social media efforts.

TRACKING NEWS COVERAGE

If you're rolling out news releases, placing releases on EurekAlert!, or pitching to reporters, you need to know whether your efforts are making a difference. By paying attention to what works and what doesn't, you can get better at your job. But perhaps more importantly, media hits tell your supervisor—and the researchers you work with—whether you're doing a good job meeting your institution's communication goals.

Paid Services

There are a number of companies you can pay to track "media placements" (i.e., news stories). These media search firms provide their clients with lists of all the news stories that mention your institution by name, including various forms of that name. For example, they may provide you with all of the stories that mention "North Carolina State University," "NC State University," or "NCSU."

You can also have these firms search for names of researchers or other terms associated with a specific project. In addition to letting you know which stories ran and where, these services also provide links to online stories, which you can then share with employers, relevant researchers, and other stakeholders. Many of these media search services will also look up media hits on your peer institutions to tell you and your employers how successful your media outreach efforts are in relation to similar research institutions.

Some of the companies that offer media search services, such as Vocus, can also be used to search for reporter contact information and to disseminate news releases. This makes them one-stop shops for your media outreach efforts. Other companies that offer media search services are more specialized, focusing on specific aspects of the media marketplace. For example, TVEyes is a media search company that targets television and radio broadcasts. They're able to focus on this niche because broadcasts are difficult to track via conventional Internet search technologies. If you want to get in-depth details on how your institution is performing in terms of TV and radio coverage, you may need to work with TVEyes or one of its competitors (such as Vocus).

These services can be incredibly useful in helping you show the institution's return on investment for media relations. In other words, they can show your employers that you are worth every penny they're paying you. What's the catch? Money. Media search services are expensive—often *very* expensive. You'll have to work with your supervisor, or whoever handles the budget for your unit, to explore the various search companies that are out there and determine whether you can afford their services.

Find Your Own Media Hits

If you can't afford a media search service, you can still track media hits yourself. I still track media hits independently, even though my institution uses a search service. Those services aren't perfect and can miss some media hits, particularly news coverage on blogs, which aren't always clearly identifiable as news outlets.

I search for media hits by using search engines, starting with Google. Here's how to search for news about your institution in general: go to Google and search for the name of your institution, in quotation marks. For example, I search for "North Carolina State University." Putting the name in quotation marks means that Google will search for the full exact phrase, rather than any combination of North, Carolina, State, and University.

When the search results come up, click on "News" so that you see only those hits that come from news outlets. Then click on "Search tools." Under Search tools, click on "Sorted by relevance" and switch it to "Sorted by date." This will bring the most recent hits to the top of the list. Then click on "Hide Duplicates" and switch it to "Show Duplicates." This will show you other stories with similar headlines. Stories that were written by the Associated Press can appear in dozens of outlets (or more), but you won't see them all unless you select the Show Duplicates option. Once you've selected all of those options, you can save this particular search by bookmarking the page. You may also want to perform similar searches for other terms, including variations on your institution's name, such as "NC State University," and bookmark those. I check these searches several times a day to keep track of breaking news regarding my institution.

I use this same approach to search for news on specific research items that I've promoted, searching for key names and terms associated with the research. If I've been pitching a story about research by Dr. Jane X on norovirus, I would search for the terms "Dr. Jane X" and "norovirus." But I would also narrow down the search by clicking on "Any time" under Search tools and switching it to "Past hour," "Past 24 hours," or "Past week." This narrows down the search considerably. However, for some reason, Google News does not pick up a lot of news stories. So it also makes sense to tap into Google's overall Web search engine.

To do this, I go to Google's landing page and type in relevant search terms, such as "norovirus" and "Dr. Jane X" (again, in quotation marks). When the search results come up, click on "Search tools" and select "Any time." Then click on "Past hour," "Past 24 hours," or "Past week." If Dr. Jane X studies norovirus, there are likely dozens or hundreds of Web pages that include both search terms. By limiting the time frame of your search, you're eliminating most of those pages, which makes it easier to spot any recent news references to Jane X and norovirus. I almost always find news stories in my Web searches that escaped the Google News search. Different search engines (e.g., Yahoo!) use different algorithms to find and sort the Web pages you're looking for, so it may be worthwhile to set up similar searches on other search engines to see if they turn up anything Google has missed.

Ask Reporters to Share

If you've corresponded with reporters about a story—pitched to them, helped them reach researchers, sent them PDFs of journal articles—you can also ask them to send you a link to their story when it goes up. Most reporters don't mind passing that along to you, and some reporters do it even if you don't ask. This is extremely useful and is one way to make sure you won't miss the article in your Web searches. This approach is also particularly helpful with broadcast stories on TV or radio, which often don't have a corresponding online story. If reporters tell you they're airing a piece, you can at least make a note of it to share with your supervisor or the relevant researchers.

SOCIAL MEDIA METRICS

Social media metrics can tell you three important things: what communication approaches lead to new followers; what kind of messages people are most likely to share; and whether you're reaching your target audiences.

Followers

This is the most basic metric for social media. How many people follow your account? By itself, this metric isn't particularly helpful; the number of followers you have doesn't tell you anything about reaching specific communication goals. However, it can serve as a useful proxy for engagement. If you're gaining followers, you are presumably providing them with engaging content; if you're losing followers, you apparently are not. Pay attention to trends. If you see a sharp spike in followers when you've been posting about the economic impact of your institution's research, it's likely a sign that your followers are interested in this kind of story. By the same token, if you see a dip in followers when you post about economic impacts, it probably means you should share content that's focused on something else.

Ultimately, social media platforms are only useful if people follow you. By the same token, having a following is only significant if it helps you accomplish your communication goals. You have to balance what your followers want with the messages you need to get across to them.

Shares, Likes, and Retweets

As with followers, shares and retweets (RTs) are metrics you can use to measure engagement with your content. People are most likely to share information that they think is interesting or important,[1] so tracking their shares and RTs provides insight into which messages are resonating with your audience.

Tracking the rise and fall of your follower numbers provides you with a general idea of how people respond to the content you're sharing. But seeing how many people have shared or "liked" a specific story, image, or video can give you a more precise idea of what people are interested in. It also gives you insight into how they want you to share information with them. For example, there are several ways you can tell people about a research finding or a grant award. You could write a blog post, create a short video, design an infographic, or superimpose brief text over a relevant photograph, to name a few options. Which way works best? Infographics or poster images using photographs may be the best fit on Facebook or Google+. Blog posts may be effective for Twitter. To find out, see which types of content your followers are engaging with—and the best metric for that is to see what *your* followers like to share with *their* followers. If your followers show a keen interest when you share blog posts, write more blog posts. If your followers like videos, invest more time in that medium. Pay attention to what your audience is telling you.

This will not only help you create more content that resonates with your followers; it will also help you amplify your message. If you're trying to get cancer researchers interested in a specific research project, you want the cancer researchers who follow you on social media to share your content with their social media followers

1. "Global 'Sharers' on Social Media Sites Seek to Share Interesting (61%), Important (43%) and Funny (43%) Things," Ipsos, last modified September 3, 2013, http://www.ipsos-na.com/news-polls/pressrelease.aspx?id=6239.

since their followers likely include a significant number of cancer researchers who don't follow you directly. In this scenario, you're not only reaching more people; you're reaching more people who are in your target audience.

It's easy to track some forms of "liking," sharing, and RTing. Facebook and Google+ both show how many people have liked or +1'd posts, as well as how many people have shared them. Similarly, Twitter tells you who has "favorited" or RT'd any of your tweets (or tweets that mention you). However, none of those platforms will tell you whether anyone has taken a link you shared and then shared it on their own account without referencing you (though Facebook does provide some limited analytics to page administrators, and Twitter has an analytics platform that is accessible to those who pay to promote their accounts). Assuming your video and blog clearly state your institutional affiliation, that's not necessarily a problem. But it does mean there's no clear way to determine who is sharing your content online, and that makes it harder to determine what audiences you're reaching.

Audience Identification

While it's important to know whether people are engaging with your content, it's just as important to know if you're reaching people in your target audiences. There are quite a few tools that can help you determine who is sharing links that you post on social media, including URL-shortening tools, which I talk about in the section on blog metrics. But two tools that I find particularly useful are Topsy and Who Shared My Link.

Topsy offers a number of analytical tools to paying customers, but it also includes a free tool that anyone can use. If you plug a URL, or Web address, into the search bar on Topsy's landing page, it tells you everyone who has shared that link on Twitter and the associated message that went with each share. If User X shared a link, it would show the full message they tweeted: "This is the greatest research story ever! [link]." This is extremely useful because it shows you who was sharing your content and gives you an idea of what they thought about it. Or with my earlier example about trying to reach cancer researchers, you could use Topsy to see whether cancer researchers have been sharing your link. If so, you know that you're reaching the audience you want. By looking at the language they used when tweeting about the link, you can tell how enthusiastic they are about the content. If they say something is "fascinating" or "important," you're doing well. But if they're raising significant questions about the quality of the research, you may want to talk to the relevant researchers. Forewarned is forearmed.

Who Shared My Link is an analytical tool provided by Muck Rack. Like Topsy, it can give you a great deal of information if you can afford to pay for the service, but there is also a free component of the tool that can provide valuable information. If you plug a URL into Who Shared My Link but you're not a paying subscriber, the site will not tell you exactly who shared your link. However, it will tell how many users have shared your link on Twitter, Facebook, Google+, LinkedIn, and a variety of other platforms. This metric is imprecise, but it gives you a good idea of how many people are engaging with your content. It also lets you know which platforms are effective for sharing your content. So if very few people are sharing your content on Facebook, you may want to revisit your strategy for that platform or abandon it altogether.

Who Shared My Link will also tell you whether any journalists who are registered with Muck Rack have shared that particular URL. It won't let you know precisely *which* journalists (you have to pay for that option), but it's still useful information because it tells you that reporters are interested in the content you're sharing. And journalists should always be part of your target audience.

BLOG METRICS

Blog metrics measure everything from the number of people who have clicked on a particular blog post to the amount of time they've spent on that page. Many blogging platforms, such as WordPress, offer a variety of "plug-in" analytic tools that give users access to a range of metrics. Bloggers can also use Google Analytics, Piwik, or other third-party software to get access to analytics data. Depending on the platform and the amount of data you want to collect, these plug-ins and third-party analytic tools are inexpensive or free, so it's worth taking the time to incorporate them into your blog. This section offers an overview of some of the metrics you can look at once you have those analytic tools in place—and how you might be able to use them.

Influence Metrics

There are a number of sites that offer various ways of measuring your online influence by giving you a score of some kind. These include Klout, Kred, and PeerIndex. These "social scoring" sites exist because social media are big business, and marketing professionals (among others) want to know who has influence and who doesn't.

That said, all of these social scoring sites have been the subject of substantive criticism on issues ranging from privacy concerns to the fact that people can try to "game" social scoring algorithms to create an inflated score.* While these social scoring sites may evolve into useful tools, it's too early to place much faith in them. I recommend against making any communication or social media decisions based on information from social scoring sites.

*"What's the Score?: The Ultimate Guide to Social Scoring," *Fliptop*, accessed November 7, 2013, http://www.fliptop.com/socialscore/.

Conventional Metrics

There are a number of metrics that most analytic tools provide, but you need to know what you're looking at. I'll explain some of the more common terms because although some seem self-explanatory, that's not true for all of them.

Visitors or *Visits:* the number of times that users have visited your overall blog. This is *not* the total number of people who have visited your blog.

Unique Visitors: This is the total number of people who have visited your blog. This number is usually less than (and never more than) the number of visits, because if the same person visits your blog three times, that counts as three visits but only one unique visitor. It is always good to get more unique visitors and expand your audience, but ideally you want significantly more visits than unique visitors, because that indicates that readers like your content enough to visit repeatedly.

Pageviews: The total number of times the pages on your blog have been visited. This is different from visits because a reader may read three different posts on your blog on the same visit, clicking from one story to the next. That would show up on your analytics as one visit but three pageviews. It's good to have more pageviews than visits, because that indicates that readers are sufficiently interested in your content to check out additional posts.

Duration or *Length of Visit:* How long did visitors spend on each page? Generally speaking, longer visits are better. It means that visitors are taking the time to read what you wrote.

Bounce Rate and *Exit Rate:* Bounce rate represents the percentage of visitors who view only one page and then exit the site, rather than viewing other pages on your site. A low bounce rate is good, because it tells you that readers are viewing multiple pages on each visit. You may also see numbers listed under "exit rate." Exit rate shows the total number of people who leave your site from a given page, including people who had previously viewed other pages on the same visit.

You can get data on any of these metrics for your entire blog—combining all of the people who have clicked anywhere on your blog—or for any single post. Viewing metrics for a single post is useful because although looking at the numbers for the blog may give you a big-picture idea of how successful you are, it can be misleading. For example, if you look at the overall numbers for your blog and see that you've gotten 20,000 unique visitors, you might think you're doing great. But when you look at how successful each individual post has been, you might find that 18,000 of those unique visitors only looked at one post and that the rest of your posts have a fairly limited readership. This can offer you insights into which subjects your readers are particularly interested in. Thus, if there are topics that are particularly appealing to your readers, it would make sense to spend more time writing about them.

You can also control the time frame that you examine with these analytic tools by looking at overarching trends over the past year or at metrics over the course of a single day. This can help you determine the best times to post content on your blog or to share it via other social media. If you find that your blog gets relatively few readers over the weekend, it would make sense for you to avoid posting new content on Saturday or Sunday. Similarly, if you discover that most of your readers visit the blog early in the day, you should post new material in the morning.

You can also track "traffic sources," which tells you exactly how visitors are arriving at your site. To do this in Google Analytics, you click on the URL for a specific post, which calls up all of the information for that page. Then click on "Secondary dimension," which opens a drop-down menu. In that menu, click on "Acquisition," which opens another menu. Select "Source" and you'll see how people found your blog post. Piwik offers similar information to users. Traffic source data can tell you whether people are coming to your blog via Twitter, Facebook, reddit, or some other site, and whether another blogger or reporter has linked to your site. You can use these data to further fine-tune your efforts to disseminate the blog. Posting new content at the right time and pushing it out through the right channels is a great way to maximize your blog's impact.

URL Shorteners

You can collect even more data on who is reading your blog—and how they're finding it—by using URL-shortening services, such as bitly. To use these free services, you create an account and plug in the URL for one of your posts. The service then offers you a much shorter URL that people can click on to reach the post. The real benefit of most URL-shortening services is that they also collect analytic data. Some services will let you know not only how many times people have clicked on a link but when they clicked on it. They can also tell you what platform people were using when they clicked the link (e.g., on Twitter), where those people are located (e.g., the United States or the United Kingdom), and even who else has shared that link on Twitter.

These metrics can give you a lot information about your audience. For example, you might learn that many of your readers are in the United Kingdom and that they look at your posts in the morning. If you are based in the United States, then you might want to schedule your posts to go up much earlier in the day, to account for the time difference between, say, New York and London. Much like Topsy, URL-shortening services can also tell you whether you're making headway toward reaching your target audiences. If you want to reach an audience of scientists in a particular field, you can check to see whether those scientists are sharing your posts on Twitter.

UNCONVENTIONAL METRICS

All of the metrics discussed above are useful, but their primary function is to help you figure out how many people you're reaching, who those people are, and how you're reaching them. However, those metrics may not tell you whether you're making progress toward achieving specific goals. To do that, you'll need to come up with your own "unconventional" metrics. For example, if your goal is to recruit people to participate in a citizen science study, a good metric would be the number of people who have signed up.

Crafting Unconventional Metrics

When it comes to unconventional metrics, there is no predetermined list of tools available. That's because unconventional metrics are dependent on the goals you've set for your blogging, pitching, and social media efforts. To find metrics that fit your needs, you have to be creative. There is no one-size-fits-all approach that works, so I'll use my institution's research blog as a short case study on how to come up with unconventional metrics. Part of my job as a university PIO is to maintain our institution's research blog. The university has goals for the research blog, but those goals do not include reaching as many readers as possible. So what are the goals for our blog, and how are we measuring progress toward them?

Goal #1: Promote university research to external audiences, such as potential students and faculty.

Example metric: Did the post help us amplify the story via external news media? If we were able to use the post as an effective pitching tool to reach reporters, then we would ultimately be able to count the stories that we got out of it. And if a story ran in a major news outlet, such as the *Los Angeles Times*, we know that we reached a fairly broad audience. Because our blog posts have led to hundreds of news articles in outlets ranging from *Scientific American* to *USA Today*, we know we are achieving this goal.

This is a good example of marrying conventional and unconventional metrics. Blogs are not often thought of as media outreach tools. Instead, people often measure the success of the blog by looking at the number of visitors it receives. I think visitors are great, but I'm under no illusions about the likelihood of our university blog becoming a go-to news source for a national audience (it's not going to happen). But I do want to use the blog to reach reporters who do have a national audience. It's an unconventional way of measuring blog "success," but it uses conventional metrics: media hits.

Goal #2: Help researchers disseminate findings (which makes funding agencies happy, can boost citation rates, etc.).

Example metric: Look at whether blog posts are being featured on the sites of funding agencies. For example, in a one-year period, 39 out of 110 blog posts were featured on the National Science Foundation's news site, Science360. That means approximately 34 percent of our posts were being highlighted by an important federal funding agency, which indicates progress toward our goal. Tracking citation rates is a great metric, but it takes years to get an accurate assessment of citation impacts—and I can't afford to wait that long to gauge success.

This is an unconventional goal that is easy to track and that resonates with most research institution administrators. Federal funding is essential to research operations at most institutions, and clear indicators of interest and support from relevant federal agencies will always be welcomed by your institution's management team.

Goal #3: Achieve specific communication goals as they arise.

Example metric: Sometimes a researcher will come to us with a specific problem or challenge and ask for our help. Because the challenge is different each time, you have to come up with a different metric for each set of circumstances.

For example, in spring 2012 a computer science researcher at my institution launched a data-sharing initiative called the Android Malware Genome Project. Data-sharing initiatives present a circular problem from a storytelling perspective: the researcher can't do any work on the initiative until people know about it and start sharing data, but there's nothing to tell people about until the researcher has started doing work. How do you spread the word when there's not much to say? Answer: short blog post. We put a four-paragraph post about the project online and did a Google search for the term "Android Malware Genome Project" that morning. It turned up only three responses, indicating that there was very little public awareness of the project. We pushed the post out via social media and online forums, and we pitched the story to a handful of reporters we thought might be interested. Three weeks later, a Google search turned up more than 300,000 hits. The word was out, and the researcher was fielding requests for additional information on the project from potential collaborators. It was a very specific communications goal, and the metric for success was the number of potential collaborators who contacted the researcher.

There is no formula that will always give you the right metrics. Spend some time and mental energy to develop meaningful ways to determine whether you're achieving your communication goals. Critical thinking is key, and don't be afraid to be creative.

Following Up with Researchers

The researchers you work with make up one of your most important audiences, so you need to find out what effect your communication efforts are having on their work. An easy and effective way to do that is to simply ask them. This should be a common practice, rather than an unconventional metric, but it's not.

Once a year I send an e-mail to all the researchers I've worked with over the past eighteen months, asking them for feedback on what it was like to work with me. I want to know whether they have any concerns or if it was a relatively painless process for them. I also ask whether our communication efforts led to collaborations with other researchers, helped them attract graduate students, earned them goodwill from funding agencies, led to licensing agreements with private companies, or benefited them in any other way. (All of those examples are things that have actually happened, by the way.) These follow-up efforts have made me a more effective PIO and have shaped a lot of the material in this book. It was through feedback from researchers that I learned the importance of pitching stories to small, discipline-specific news outlets.

Feedback from researchers can also be surprising. There have been instances where a news release or other communication effort that I thought was a failure turned out to be a significant success.

For example, a news release may have led to only a single story, which does not look good in the metrics column. But that release also caught the attention of a researcher at another institution, who followed up with the relevant researcher and became a key collaborator in subsequent research.

You can't buy software to assess these metrics, but they are enormously important as means of gauging your success and demonstrating the value of science communication to your institution.

KEEP TRACK AND SHARE YOUR METRICS

Whatever metrics you are using, it is essential to keep a record of how you are doing. If your supervisor wants to know what you've done over the past year, you need to defend your performance with real numbers. When it comes to assessing your success as a PIO, your managers shouldn't have to take your word for it. Similarly, if you've had a particularly noteworthy success, let people know. You shouldn't be a smug braggart, but you should send an e-mail to your management team outlining your achievement. It can be as simple as saying, "I wanted to make sure you saw this. I'm really pleased with how it turned out." If you don't share your successes, no one else will.

KEY POINTS

- Track your news coverage, using paid services or search engines.
- Followers, shares, likes, and retweets can help you identify the content that your audiences find most engaging.
- Use online tools, such as Topsy, to determine whether your social media efforts are reaching your target audiences.
- Conventional metrics can help you determine how to use blogs effectively.
- Unconventional metrics take more work, but they give you keener insights into the impact of your communication efforts.
- Follow up with researchers to see if your communication efforts are making a difference in their work.
- Keep track of your metrics, and share your success stories.

STORIES YOU DON'T WANT
CRISIS COMMUNICATIONS

Researchers and research institutions can come under fire for a variety of reasons, ranging from ethical violations by researchers to political attacks against climate-change research to protests by animal rights organizations. This chapter will focus on how to handle bad news by being prepared, responsive, and honest. This chapter also discusses communication plans. These are an essential element of crisis communications, but they're also useful in other circumstances for audience identification and messaging.

BE PREPARED

There are two types of crises: those that are generated within your institution and those that are caused by someone outside of it. Examples of internally generated crises include misuse of research funds and researchers who falsify data; externally generated crises include politically motivated attacks against your institution's research or hackers breaking into your computer system and accessing personal information about your staff. Because these crises can be caused by a single bad actor in your organization or by an outside group, it can be difficult or impossible to foresee a potential crisis situation. But you can still take steps to be prepared—and the more prepared you are, the better you'll be able to handle the crisis.

Identifying Key People

In the event of a crisis, it is essential to share reliable information as quickly as possible so that the proper people can make informed decisions. For example, if there's been a hazardous chemical spill at your institution, you need to tell your employees how to respond. Should they stay in their offices? Should they leave immediately?

Should they not come to work the next day? In a situation like this one, sharing information isn't a luxury—it's a responsibility.

The emphasis here is on sharing *reliable* information. Sharing inaccurate or unconfirmed information will only contribute to confusion. It can also get you into a lot of trouble. So it's crucial for you to identify the people who will be able to provide important information and make key decisions in the event of a crisis. If you know who those people are ahead of time, you'll be able to act quickly if something bad happens. If someone breaks in to a facility at your institution and releases laboratory animals, you should know who can answer questions about your institution's security, about the risk (if any) that those animals pose to public health, and about what this means for ongoing research at your institution. Keep in mind, of course, that for security reasons you may not be able to disclose all that you know in these situations. For instance, researchers are sometimes the target of threats from animal rights activists. You wouldn't want to do anything that could make those researchers more vulnerable to attack. In short, you need to know where to turn to get reliable information about what happened, how it happened, and what will happen next.

You also need to know who will be speaking on behalf of your institution. Crises can be complicated, and you should have one or two spokespeople who will be able to pull together all of the relevant information and explain the situation to reporters. This is usually a high-ranking administrator who can represent the institution and answer questions from reporters.

Know How to Reach Key Stakeholders

As a PIO, everything you do on a daily basis involves reaching out and building relationships with the people you'll need most in a crisis. In addition to talking to reporters, you'll need to communicate directly with key stakeholders. For example, if there is a chemical spill on a university campus, the university needs to be able to quickly reach faculty, staff, and students (and their parents) with essential safety information. Similarly, if you discover that a key administrator at your institution has engaged in large-scale scientific fraud, you'll need to reach your institution's board, donors, relevant federal agencies, and other key audiences.

You should not have to figure out who your key audiences are or how to reach them on short notice. Your institution needs to have a plan in place for how to reach these audiences quickly in the event of a crisis. Also keep in mind that during a crisis, key audiences aren't always educated audiences. In a large-scale calamity, the media will throw all of its resources into covering the story. The reporters covering your crisis may be from the political desk or the sports page. They may not know the first thing about who you are, what you do, or how you work. Despite this, their report may be the leading story on the evening news. For this key group of people, it helps to have quick and easy-to-digest materials explaining the basics about your institution and what they need to know. This often helps reporters frame their stories and ensures that they'll come back to you if they need to follow up.

Again: be prepared.

Identify Potential Problem Areas

While you shouldn't expect bad things to happen, you should know that they can happen. If there are particular areas that you think could be targets of protests or sabotage, you should plan accordingly. So if your institution engages in climate-change research, you should plan in advance how to deal with politically motivated attacks from politicians or organizations who are vocal opponents of this research. Similarly, if your institution uses laboratory animals to study toxicology or human health, you should be prepared to answer questions or deal with attacks from animal rights organizations.

After you identify any areas that could serve as crisis focal points, come up with a rough communication plan ahead of time. (More on communication plans later in this chapter.) A great way to identify problem areas is to conduct crisis communications drills and scenario planning. The best scenarios are drawn from events that have happened in the past and get thrown together in a perfect storm of activity. These scenarios are designed to challenge you and your colleagues. They force you to make decisions with incomplete information and inadequate resources. For instance, how would you react if you had to deal with a security breach, a computer hacking, and an allegation of sexual harassment—all on the same day? How do you get the word out if someone has taken over your social media accounts? How would you handle it if you suddenly discovered that one of your researchers has been providing false information he got from a dishonest colleague? What do you do if you wake up in the morning to discover a controversy that went viral on social media a few hours ago? And how would you know if it's a hoax? These are all scenarios that have happened at one time or another in a wide variety of organizations. Don't think they can't happen to you.

As you review your performance after a drill, you can identify the places where you have gaps in accountability or knowledge before a real crisis happens.

BE HONEST

The single most important thing to remember in crisis communications, especially if someone at your institution is at fault, is to be honest. I once asked a crisis communication specialist what the key was to handling crisis communication efforts effectively. His response was succinct, so I will repeat it verbatim: "Own your shit." Do not try to cover it up. Do not lie about it. Do not try to explain why "it really wasn't that bad." All of those things will just make it worse. And doing nothing is just as bad as being disingenuous. This is called the "bunker mentality," where you hunker down, don't say anything, and hope the problem goes away. It never goes away. Instead, the bunker mentality makes you seem helpless (at best) or out of touch and unrepentant (at worst).

If your organization has done something wrong, admit it. In full. Don't let reporters drag it out of you bit by bit, feeding a lengthy series of stories about what you did wrong and how you tried to hide it. Get it all out there. You will take your hits, but you'll take them all at once, and you'll be in a better position to move on. This is sometimes easier said than done. Organizations aren't monolithic; colleagues will often have different perspectives on what the truth really is, who is responsible, or what to do. Human resources policies are in place for a reason, and people often have legitimate privacy rights on the job. Finally, admitting fault is also sometimes the precursor to paying out a lot of money.

All that said, a complete admission shortens the length of a story. It also demonstrates to your stakeholders that you "get it"—and since you understand the problem, you're in a position to fix it.

BE PROACTIVE

Being honest is essential, but it's not enough. If your institution, or someone at your institution, has done something wrong, you need to be proactive about addressing the problem.

Acknowledge the Problem

The first step is to acknowledge that the problem exists. I don't mean that you should acknowledge that someone *claims* there is a problem; I mean that you have to acknowledge that there *is* a problem. And you have to explain exactly what the problem is, so that everyone knows that you understand it. This also ensures that everyone is talking about the same thing.

Messaging is critical here. Your institution needs to speak in a clear and decisive tone. That means you should avoid the passive voice that you hear in statements such as, "Mistakes were made." I noted earlier in this chapter that you need to identify your key spokespeople, and that preparation comes into play here. The messenger matters. For example, if someone has been gravely injured or killed, the head of the organization should be front and center. This demonstrates that the organization is taking the incident seriously at the highest levels.

Apologize

If you've done something wrong, you need to apologize fully, openly, and honestly to all affected parties, as well as any key stakeholders. A crisis can rattle the public's faith in your institution. Apologizing is an essential step toward rebuilding that trust.

Assess

You need to understand how this problem happened in the first place. For instance, if it was a chemical spill, what went wrong with your safety protocols? If it was researcher misconduct, how was he or she able to get away with it? To help determine what went wrong, your organization should consider bringing in trusted third parties to assess the situation and suggest next steps. They may spot mistakes that your institution hadn't noticed. This also indicates that your institution is serious about addressing the circumstances that led to the crisis in the first place. This is where planning can come

Communicators and Lawyers

This chapter's section on being proactive in crisis communications is a wish list. It's how I wish people would handle crisis communications, and it's the counsel that I would give to anyone who asked. However, you should know that it may not run that smoothly.

Lawyers protect an organization from liability. PIOs protect that organization's reputation. These are different jobs, requiring different strategies and tactics. They will come into conflict at times. When conflicts arise, it is the organization's administration that makes the final call on which way to go. For example, your institution's attorneys may oppose public acknowledgment of a problem or making a public apology, out of concerns that these acts could open your institution to legal attacks and liability. If you're a PIO, I urge you to tell your institution to be as open and honest as possible. But ultimately you will have to live with whatever decisions are made by your institution's administrators. Do not speak out on behalf of the institution without your institution's permission.

into play. In a crisis, it can be difficult to identify trusted third parties and establish communications with them. Ideally, you'll know who those third parties are ahead of time, and you'll have already put yourself in a position to work with them.

Outline Your Response

Once you've established what went wrong and why, you should publicly outline how your institution is going to deal with the problem to prevent future mistakes. You should also outline how your institution is doing to deal with the consequences of what has already happened. You cannot undo a past mistake, but you can take steps to ameliorate the consequences of that mistake.

BE RESPONSIVE

As I noted earlier in this chapter, crises are not always your institution's fault. Regardless of who was at fault, your organization should always be responsive. Respond to questions from reporters or other stakeholders quickly, clearly, and accurately. If you are still waiting to confirm relevant facts, say so. People are more likely to be patient if you tell them that you are trying to get good information than if their questions are met with silence. In the social media age, sometimes "quickly" means "now." Once you have responses drafted, it may make sense to have someone monitoring social media channels to provide your message directly to anyone who discusses it online. Sometimes all it takes is one retweet to spin things in the wrong direction and hopelessly out of control.

COMMUNICATION PLANS

A communication plan is a document that establishes steps for how to communicate about a specific subject. It serves three goals. First, it makes you think through the process, enabling you to identify any questions or problems you might not otherwise think of. Second, it can serve as a reference document, to help you make sure you're not forgetting anything important in the event of a crisis, when it can be easy to lose your train of thought. Third, a clearly written communication plan lets everyone involved know what they're supposed to be doing and makes sure that everyone is on the same page.

Most PIOs won't have to develop communication plans because they are often written by management-level communication officials. However, I think it's important for PIOs to have at least a rudimentary overview of communication planning, because some of you *are* managers—or will be. Even if you don't have to develop your own communication plans, this may help you understand the plans you're being asked to help implement.

Institutions develop communication plans for a wide variety of reasons, not just for crisis communication purposes. An institution may write a communication plan to attract new researchers. I'm including communication plans in the chapter on crisis communication because planning can significantly improve your institution's response to a crisis.

Most communication plans include the same components. I'll explain each of those elements in turn, using the example of a medical lab that uses animal research.

Introduction/Background

This should be a concise overview of the relevant issue. At a medical research institution that is concerned about possible protests against its animal research facilities, you would want to outline what sort of animal research your institution engages in and why it's important.

Objectives

This should clearly articulate what you hope to accomplish through your communication efforts. For instance, you may want to raise awareness of the role that animal testing plays in protecting public health and developing new drugs or other medical advances.

Audiences

This should specify who you are hoping to reach. Key audiences may include your institution's employees, prominent local and national news outlets, or federal agencies that fund your research initiatives. Keep in mind that each audience has its own influential leaders. If you know those leaders and have relationships with them, they can help you get information to a large community quickly when necessary.

Key Messages

Your key messages are the points that you want to convey to your key audiences. These will also likely serve as talking points for your institution's administrators or spokespeople. For a medical lab, your key points might be that you are in compliance with all existing best practices for the ethical treatment of laboratory animals and that you are engaged in research that is improving our understanding of cancer, Parkinson's disease, or other illnesses.

These key messages may evolve quickly in the event of a crisis as new information comes in, so be prepared to make changes. You may want to develop these key messages in partnership with your general counsel's office.

Tactics

This section should lay out specific steps you will take to make progress toward your goals. You may want to have internal seminars or use other internal communication channels to brief employees on your animal research efforts. You could also invite reporters to interview administrators and researchers or to tour research facilities. You may want to have alternate social media channels or an emergency website you can fall back on in the event of an ongoing crisis. An emergency website is essentially a stripped-down version of your homepage, providing key information that can be updated as needed. The simplified nature of the site will help keep your servers from crashing if you receive a large number of visitors seeking information during a crisis.

Timeline

This isn't absolutely necessary, but a timeline can help you decide when you need to implement your communication tactics. In a crisis situation, a timeline may quickly go out the window. But if you're proactively planning for a specific goal, such as recruiting new researchers, you could lay out a schedule for pitching specific stories to relevant news outlets, publishing relevant research stories on your blog or institution homepage, and making a concerted social media push. If you get the timing right, those efforts could raise your profile significantly with your target audiences.

KEY POINTS

- Know who your relevant experts and spokespeople are *before* you need them.
- Know who your key audiences are and how to reach them.
- Don't be a pessimist, but keep an eye out for potential problem areas ahead of time, and plan accordingly.
- Be honest. Any attempt to ignore a problem or talk your way out of it will only make the situation worse—much worse.
- If your institution did something wrong, you need to acknowledge the problem, apologize, and publicly outline the steps you'll take to make sure it doesn't happen again.
- Be responsive. Share information and answer questions accurately and as quickly as possible.
- Communication plans are *not* just for crisis communications, but they are enormously useful in a crisis.

CONCLUSION

The Science PIO Commandments

Being a science PIO can be extremely rewarding, particularly for anyone who is fascinated by science and technology. Science PIOs are often among the first people to learn about—and write about—new research findings. That's exciting. And PIOs also get to work with both researchers and reporters, often developing strong relationships (even friendships) that last for years. But those strong working relationships require trust, which can be difficult to earn and easy to lose.

PIOs often have a bad reputation among reporters, and not without reason. Bad PIOs can be annoying, misleading, frustrating, and whatever the opposite of helpful is. When I made the move from reporter to PIO, I created a list of commandments for myself. Some of these are specific to science PIOs, but most apply to everyone in the business. These commandments are important if you want to establish trust and protect your working relationships. If some sound familiar, it's because I've mentioned them earlier in this book.

DO NOT OVERSTATE FINDINGS

When you are writing about research findings, you should explain them. You should place them in context. You should not blow them out of proportion or overestimate the researcher's findings until they have little relation to reality. Don't say there will be flying cars or a cure for cancer unless the researcher has built a flying car or developed the cure for cancer (and if it is the cure for cancer, tell us precisely which form of cancer has been cured).

RESPOND PROMPTLY

Reporters are busy people who are trying to meet deadlines. If you get a phone call or an e-mail from a reporter, respond as quickly as possible. Reporters notice when they get a prompt response, and they appreciate it. Reporters also notice when it takes forever for a PIO to get back to them. They do not like that.

And remember, if you don't have the information a reporter wants, respond right away and say that. At least the reporter won't be wondering if you can provide a missing piece of the story—and can start looking for that piece somewhere else.

HELP OTHER INSTITUTIONS (AND DON'T BAD-MOUTH THEM)

When a reporter contacts you looking for an expert, try to help. If your employer doesn't have a relevant expert, send them someplace else. I work for North Carolina State University, which does not have a law school or a medical school (for humans, anyway—if you're an animal, our veterinary medicine school has you covered). When reporters contact me with questions best suited to law or medical professors, I always refer them to our neighbors at Duke or UNC–Chapel Hill. This is good for business in two ways. First, it helps the reporter find the source he or she is looking for—even though that source isn't at your institution. Reporters appreciate this and will remember it. Second, it fosters good relationships with other institutions—which might be able to return the favor at some point.

Do not insult or put down other institutions. Kind words can help you. Being a jerk never pays off.

DO NOT BLOCK ACCESS TO RESEARCHERS

Reporters want to talk to whoever did the research. They do not want to talk to you. It is your job to facilitate access to the researchers. I usually give reporters the direct contact information for the relevant researchers, so they do not need to call me or interact with me in any way if they don't want to. The easier I make their job, the more likely it is that they'll write about the research. Also, it makes it more likely that they'll be willing to read my e-mails in the future.

I know that not all institutions allow direct access to the researchers. I feel fortunate to work at an institution that does.

FACT-CHECK YOUR NEWS RELEASE/ BLOG POST WITH THE RESEARCHERS

If you are writing a news release or blog post about research findings, you should run a draft of the release or post by the relevant researchers or at least the lead researcher. Reporters do not want to be misled because you got your facts wrong. And researchers do not want their work misrepresented to the public. Any mistakes you make may be unintentional, but that will not be very comforting if reporters and researchers think you've been stupid (at best) or disingenuous (at worst).

BE ABLE TO PROVIDE THE PAPER

You might be the greatest news release writer in the history of news release writers, and you might do a wonderful job of summing up the key findings in a new journal article, but reporters are not going to take your word for it. They will want to see the actual paper.

If the paper is available on open access, put a hyperlink to the journal article in your release or blog post so that readers can go directly there if they want to. If the paper is not open access, put the hyperlink in anyway. Some reporters may have a relevant subscription, as will many researchers or other parties who are interested in the release. But always make sure you have a PDF of the article available that you can send to interested reporters.

ACKNOWLEDGMENTS

I'd like to thank Al Sosenko for giving me my first opportunity as a reporter, even though I had no experience; and Keith Nichols for giving me my first opportunity as a PIO, even though I had no experience. If they hadn't given me a chance, I have no clue what I'd be doing with my English degree right now.

I'd also like to offer particular thanks to Mick Kulikowski and Brent Winter. Mick has been my mentor as a PIO. If I'm good at this job, it's largely due to his guidance. And Brent is as good an editor as any writer could hope for. Any mistakes in this book are mine alone and come despite their best efforts.

People who have offered advice and encouragement, whether they knew it or not: Melissa Anley-Mills, Jeremy Bernstein, Deborah Blum, Tom Breen, Marty Coyne, Erin DeWitt, Nadia Drake, Rob Dunn, Adrian Ebsary, Rose Eveleth, D'Lyn Ford, Fred Hartman, Christie Henry, Scott Huler, David Hunt, Amanda Moon Johnson, Mary Laur, Glendon Mellow, Holly Menninger, Melinda Wenner Moyer, Ivan Oransky, Tracey Peake, Tim Peeler, Rob Perez, Eleanor Spicer Rice, Jimmy Ryals, Paul Singer (I quote your editing advice almost daily), Logan Ryan Smith, John Stanton, Janet Stemwedel, Gisela Telis, Shawn Troxler, David Wescott, Alex Wild, Clifton Williams, Emily Willingham, Carl Zimmer, and Anton Zuiker.

And, of course, none of this would be possible without the support of Julia Ellis, Karen Roeseler Shipman, Nora Shipman, Fiona Shipman, and Violet Shipman. Thank you.

APPENDIX A USEFUL LINKS FOR MULTIMEDIA

As with all things online, these sites may change or disappear altogether. But this should give you a good idea of available resources.

GOVERNMENT IMAGE SITES

CDC Public Health Image Library: http://phil.cdc.gov/phil/home.asp
Department of Energy Photography:
 http://www.flickr.com/photos/departmentofenergy/sets
NASA Multimedia:
 http://www.nasa.gov/multimedia/imagegallery/index.html
National Renewable Energy Laboratory Image Gallery:
 http://images.nrel.gov/
National Science Foundation Multimedia Gallery:
 http://www.nsf.gov/news/mmg/
NIH Images from the History of Medicine:
 http://www.nlm.nih.gov/hmd/ihm/
NOAA Photo Library: http://www.photolib.noaa.gov/
U.S. Geological Survey Photographic Library: http://libraryphoto.cr.usgs.gov/

OTHER FREE STOCK IMAGE SITES

Flickr: http://www.flickr.com/
freeimages: http://www.freeimages.com/
FreePixels: http://www.freepixels.com/
Wikimedia Commons: http://commons.wikimedia.org/wiki/Main_Page

FEE-BASED STOCK IMAGE SITES

Corbis: http://www.corbisimages.com/
Dreamstime: http://www.dreamstime.com/
iStock: http://www.istockphoto.com/
Shutterstock: http://www.shutterstock.com/

GOOD SITES FOR BACKGROUND ON FAIR USE AND RELEASE AGREEMENTS

American Society of Media Photographers:
 http://asmp.org/tutorials/property-and-model-releases.html
Columbia University Fair Use Checklist:
 http://copyright.columbia.edu/copyright/fair-use/fair-use-checklist/

ONLINE MULTIMEDIA TUTORIAL SITES

Knight Digital Media Center, Berkeley:
http://multimedia.journalism.berkeley.edu/tutorials/
Lynda.com: http://www.lynda.com/
New York Video School: http://www.nyvs.com/
Poynter News University: http://www.newsu.org/
Video Copilot: http://www.videocopilot.net/

APPENDIX B SAMPLE RESEARCH NEWS RELEASE

INJECTABLE "SMART SPONGE" HOLDS PROMISE FOR CONTROLLED DRUG DELIVERY

Media Contacts:
Zhen Gu, (919) 515-7944, zgu3@ncsu.edu
Matt Shipman, NC State News Services, (919) 515-6386, matt_shipman@ncsu.edu

For Immediate Release
Researchers have developed a drug delivery technique for diabetes treatment in which a sponge-like material surrounds an insulin core. The sponge expands and contracts in response to blood-sugar levels to release insulin as needed. The technique is in the early stages of development but may eventually also be used for targeted drug delivery to cancer cells.

"We wanted to mimic the function of healthy beta-cells, which produce insulin and control its release in a healthy body," says Dr. Zhen Gu, lead author of a paper describing the work and an assistant professor in the joint biomedical engineering program at North Carolina State University and the University of North Carolina at Chapel Hill. "But what we've found also holds promise for smart drug delivery targeting cancer or other diseases." The research team includes Daniel Anderson, the senior author and an associate professor of chemical engineering and member of the Koch Institute for Integrative Cancer Research at MIT, and researchers from the Department of Anesthesiology at Boston Children's Hospital.

The researchers created a spherical, sponge-like matrix out of chitosan, a material found in shrimp and crab shells. Scattered throughout this matrix are smaller nanocapsules made of a porous polymer that contain glucose oxidase or catalase enzymes. The sponge-like matrix surrounds a reservoir that contains insulin. The entire matrix sphere is approximately 250 micrometers in diameter and can be injected into a patient.

When a diabetic patient's blood sugar rises, the glucose triggers a reaction that causes the nanocapsules' enzymes to release hydrogen ions. Those ions bind to the molecular strands of the chitosan sponge, giving them a positive charge. The positively charged chitosan strands then push away from each other, creating larger gaps in the sponge's pores that allow the insulin to escape into the bloodstream. In type 1 and advanced type 2 diabetes, the body needs injections of insulin, a hormone that transports glucose—or blood sugar—from the bloodstream into the body's cells.

As the insulin is released, the body's glucose levels begin to drop. This causes the chitosan to lose its positive charge, and the strands begin to come back together. This shrinks the size of the pores in the sponge, trapping the remaining insulin.

While this work created hydrogen ions by using enzymes that are responsive to glucose, the technique could be simplified to target cancers by eliminating the enzymes altogether. Tumors are acidic environments that have high concentrations of hydrogen ions. If the sponge reservoir were filled with anticancer drugs, the drugs would be released when the chitosan came into contact with the hydrogen ions in tumor tissues or cancer cells.

"We can also adjust the size of the overall 'sponge' matrix as needed, as small as 100 nanometers," Gu says. "And the chitosan itself can be absorbed by the body, so there are no long-term health effects."

In tests using diabetic laboratory mice, the researchers found the sponge matrix was effective at reducing blood sugar for up to 48 hours.

"We've learned a lot from the 'sponge' research and will further optimize it. Meanwhile, we are already exploring applications to combat cancer," Gu says.

The paper, "Glucose-Responsive Microgels Integrated with Enzyme Nanocapsules for Closed-Loop Insulin Delivery," is published online in *ACS Nano*. The research was supported by a grant from the Leona M. and Harry B. Helmsley Charitable Trust Foundation, and a gift from the Tayebati Family Foundation.

-shipman-

Note to Editors: The study abstract follows.

"GLUCOSE-RESPONSIVE MICROGELS INTEGRATED WITH ENZYME NANOCAPSULES FOR CLOSED-LOOP INSULIN DELIVERY"

Authors: Zhen Gu, North Carolina State University and the University of North Carolina at Chapel Hill; Tram T. Dang, Minglin Ma, Benjamin C. Tang, Yizhou Dong, Yunlong Zhang, and Daniel G. Anderson, MIT and Children's Hospital Boston; Hao Cheng and Shan Jiang, MIT

Published: July 8, *ACS Nano*

DOI: 10.1021/nn401617u

Abstract: A glucose-responsive closed-loop insulin delivery system represents the ideal treatment of type 1 diabetes mellitus. In this study, we develop uniform injectable microgels for controlled glucose-responsive release of insulin. Monodisperse microgels (256 ± 18 μm), consisting of a pH-responsive chitosan matrix, enzyme nanocapsules, and recombinant human insulin were fabricated through a one-step electrospray procedure. Glucose-specific enzymes were covalently encapsulated into the nanocapsules to improve enzymatic stability by protecting from denaturation and immunogenicity as well as to minimize loss due to diffusion from the matrix. The microgel system swelled when subjected to hyperglycemic conditions, as a result of the enzymatic conversion of glucose into gluconic acid and protonation of the chitosan network. Acting as a self-regulating valve system, microgels were adjusted to release insulin at basal release rates under normoglycemic conditions and at higher rates under hyperglycemic conditions. Finally, we demonstrated that these microgels with enzyme nanocapsules facilitate insulin release and result in a reduction of blood glucose levels in a mouse model of type 1 diabetes.

Notes for the Reader:
Note that this sample release includes direct contact information for the lead researcher, as well as describing the nature of the work and its potential applications. I also went out of my way to acknowledge the role of researchers from other institutions, both because it's the intellectually honest thing to do and because it's important to ensure that the researchers at my institution maintain good relations with their collaborators. When I issued this release, in July 2013, I included a hyperlink to the paper, a high-resolution microscopy image of the sponge-like material, and an animated gif showing how the smart sponge expands and contracts.

APPENDIX C SAMPLE GRANT ANNOUNCEMENT

RESEARCHERS SEEK TO CONTROL PROSTHETIC LEGS WITH NEURAL SIGNALS
Most people don't think about the difference between walking across the room and walking up a flight of stairs. Their brains (and their legs) automatically adjust to the new conditions. But for people using prosthetic legs, there is no automatic link between their bodies and the prosthetics that they need to negotiate their surroundings.

Researchers from North Carolina State University and the University of Houston (UH) are hoping to change that with a four-year, $1.2 million collaborative project funded by the National Science Foundation (NSF).

"Our goal is to improve mobility for people using prosthetics, lay the groundwork for a new generation of prosthetic devices, and improve our understanding of how brain signals and neuromuscular signals are coordinated," says Helen Huang, principle investigator (PI) of the NSF grant and an associate professor of biomedical engineering at NC State and UNC.

In recent years, researchers have developed powered prosthetic devices that use internal motors to improve the motion of the artificial limb. The goal of the NSF project is to improve the connection between the prosthetic and the person using it.

Huang's team will be using sensors to pick up the neuromuscular control signals from residual muscles in the area where the prosthetic is connected to its user. Huang's goal is to develop an algorithm that translates those neuromuscular signals into machine language that will control the powered prosthesis—making it easier for the user to move seamlessly from standing up, to walking across the room, to climbing the stairs.

Huang's team also plans to build a prototype power prosthesis that incorporates the new technology. This aspect of the research builds on Huang's previous experience in designing and fabricating power prosthetics.

Huang's co-PI on the project, Jose "Pepe" Contreras-Vidal of UH, will be exploring ways to use neurological signals from the brain to control prosthetic legs. This is particularly important for patients who have little or no residual muscle in the area of the missing limb, because that lack of muscle makes it difficult to pick up neuromuscular signals. In those cases, signals picked up directly from the brain may be able to control the prosthetics.

"Ultimately, we'd like to combine both approaches, using signals from the muscles and the brain to provide better control of lower-body prosthetics," Huang says.

Notes for the Reader:

I wrote this sample grant announcement as a blog post, which gave me more flexibility in how I approached the story. For example, my lede focuses on defining the problem, because I thought it was important to place the grant award in context. Note that the headline highlights the work that the grant will support, not the fact that a researcher had received a grant. The body of the post describes the work that will be done under the grant and why it's interesting and important. It also acknowledges the roles of both NSF and the University of Houston, NC State's partner under the grant.

Because it's a blog post, and not a news release, I did not incorporate a section that explicitly lists the contact information for the researcher. However, I did include hyperlinks to the websites of both researchers mentioned in the piece—making it easy for reporters to contact them. I also included contact information for the researchers when I pitched the story to reporters.

In addition to still images, this announcement also linked to a page including video of earlier prototypes of powered prosthetics Huang had developed.

APPENDIX D SAMPLE MEDIA ADVISORY

NC STATE WILL DEDICATE TERRY CENTER ANIMAL HOSPITAL MAY 6

Media Contact:

Tracey Peake, NC State News Services, (919) 515-6142, tracey_peake@ncsu.edu

Media Advisory

North Carolina State University will dedicate the Randall B. Terry, Jr. Companion Animal Veterinary Medical Center at 2 p.m. on May 6. The 110,000-square-foot addition will more than double the current size of NC State's Small Animal Veterinary Teaching Hospital (VTH), making it one of the largest, most technologically advanced veterinary facilities in the country.

Media are invited to attend the event, which will take place at the Terry Center, located at NC State's College of Veterinary Medicine on William Moore Drive. Tours of the facility are available for interested media prior to the event on Thursday, May 5, and Friday, May 6. To schedule a tour, get directions, or obtain parking passes, contact Tracey Peake at (919) 515-6142.

The new medical center will double the number of exam rooms and surgery suites previously available at the Small Animal VTH, and will add expanded patient visitation areas, dedicated teaching space, and a spacious new pharmacy.

Combined, the Terry Center, the Equine and Farm Animal Veterinary Center, and the emerging Veterinary Health and Wellness Center (housed in the Small Animal VTH) will create the Veterinary Health Complex on NC State's Centennial Biomedical Campus (CBC). The CBC advances biomedical research, bio- and agro-security, food animal health and food safety, ecosystem health, animal welfare, and the critical job of training the next generation of veterinarians and veterinary scientists.

The $72 million project was made possible, in part, by a $20 million donation from the R.B. Terry, Jr., Charitable Foundation—one of the largest private gifts ever given to NC State. The North Carolina General Assembly appropriated $38 million for the facility in 2006, and additional private funds paid for the remaining costs.

-peake-

Notes for the Reader:

A media advisory is simply a concise, event-oriented news release. My colleague Tracey Peake wrote this advisory, and it does everything you want a media advisory to do: it explains what the event is, why it might be of interest, and how it is relevant to the research mission of the institution. It also tells reporters when and where the event is being held.

When pitching a media advisory to reporters, be sure to highlight any aspects of the event that may be of particular interest—such as aspects of the event that may make for good video if the reporter works in television news.

APPENDIX E **SAMPLE NEWS TIP**

NC STATE EXPERTS CAN DISCUSS JAPAN NUCLEAR CRISIS

For Immediate Release

Media looking for information on a variety of topics surrounding the Fukushima Daiichi Nuclear Power Plant crisis can contact the following North Carolina State University experts:

Nuclear Reactor Function and Physics

Nuclear engineer Dr. Paul Turinsky can discuss how a nuclear reactor works and the attempts to prevent reactor meltdown. He can be reached at 919-515-5089 or paul_turinsky@ncsu.edu.

Dr. Ayman Hawari, professor of nuclear engineering, directs NC State's nuclear reactor program and studies the interaction of radiation with matter. He can be reached at 919-515-4598 or ayman.hawari@ncsu.edu.

Earthquakes, Tsunamis

Dr. Del Bohnenstiehl, associate professor of marine, earth, and atmospheric sciences, is an expert on marine tectonics. He can discuss earthquakes and tsunamis like the ones that struck Japan on March 11. He can be reached at 919-515-7449 or delwayne_bohnenstiehl@ncsu.edu.

Dr. Mervyn Kowalsky, professor of civil engineering, can talk about earthquake engineering design and analysis, including seismic performance and behavior of different materials. He can be reached at 919-515-7261 or kowlasky@ncsu.edu.

Dr. Karl Wegmann, assistant professor of marine, earth, and atmospheric sciences, is an expert in the field of active tectonics, the study of past earthquakes from geologic and historic records, and the ways earthquakes affect and reshape geography. He can be reached at 919-515-0380 or karl_wegmann@ncsu.edu.

Communication with the Public

Dr. Bill Kinsella, associate professor of communication, can discuss how societies communicate about nuclear power, public communication on the nuclear emergency in Japan, and potential national and international implications for nuclear policies, regulation, and economics. He can be reached at wjkinsel@ncsu.edu.

Japanese Society and Culture

Dr. David Ambaras, associate professor of history, can discuss modern Japanese society and culture, as well as Japanese experiences with previous natural disasters. He can be reached at 919-513-2228 or david_ambaras@ncsu.edu.

-30-

Notes for the Reader:

A news tip is essentially a menu of experts that reporters may use to find sources that can shed insight into current events. You can issue a news tip in response to a breaking news event (as we did in 2011 with the above example), or it can be scheduled to address upcoming news. For example, we also issue a news tip listing hurricane experts in advance of hurricane season each year.

A news tip should briefly state the subject that you're focusing on; in this case, the subject was the Fukushima disaster. The news tip should then list the name and area of expertise for each of the people you're including in the tip, as well as direct contact information for those experts. It's important to be clear on how each individual's expertise is relevant to the subject of the news tip.

When pitching a news tip, it's a good idea to reach out to editors and producers as well as reporters. Editors and producers oversee story assignments and work with multiple reporters, so they are in a position to disseminate your list of experts to reporters working on relevant stories.

APPENDIX F SAMPLE COMMUNICATION PLAN

Every communication plan is different, depending on what you are trying to accomplish, the audiences you're hoping to reach, and the tools you plan to use to achieve your goals. This is a blank communication plan that you can use as a rough template. You may want to revise the template according to your goals and the resources you have at your disposal.

DRAFT COMMUNICATIONS PLAN

Goal

Set clearly defined goals here. For example, if your institution is opening a new research facility: "Raise the profile of the John Doe Research Center (JDRC) building as a world-class facility in higher education, enhancing public knowledge and regard for Generic State University (GSU) and its research community. Show our stakeholder communities that GSU is supporting the success of our students, faculty, researchers, and partners with a transformative resource that will soon be recognized as nothing less than one of the best interdisciplinary research spaces in the country."

Background

For those less well-versed on the topic, some supporting background information may be necessary to provide context or challenges.

Audience

With whom are you trying to communicate your message? Clearly define your audience or audiences here. "Everyone" is not—and should not be—your target audience. For the JDRC, target audiences may include those inside the institution (the people who will be working in and around the center), the institution's peers (who may want to collaborate with the new center), and political stakeholders in the institution's state or region (who may have approved funding for the center and who now want to see results from the center).

Key Messages

What exactly are you trying to say to make progress toward your goals? You might have three or four key messages, depending on your goals. For example, "The JDRC will give us an edge in student and faculty recruitment and retention—and be an international destination for those who want to explore how research, learning, and interdisciplinary collaboration can be transformative catalysts for a vibrant economy and culture."

Supporting Talking Points

These aren't written for the general public, but should serve as a guide for your institution's personnel. Basically, these are the things you want your people to be saying. A talking point for our generic example might be as follows:

THE JDRC DEFINES THE COLLABORATIVE RESEARCH CENTER OF THE FUTURE. With world-class research facilities in a variety of disciplines, the JDRC has also preserved significant collaborative spaces to foster interdisciplinary networking and sharing of ideas across disciplinary boundaries. The JDRC is designed to enable experimentation, support innovative projects and partnerships, and showcase university research and scholarship.

Deliverables and Timeline

Deliverables are the specific actions you will take to make progress toward your goals, as well as any supporting materials you need to support those actions. The timeline tells you when you plan to perform each action. You'll also want to make clear who is responsible for performing each item on the timeline.

Below, you will find a template chart you can use for planning your timeline of deliverables. You'll likely want to create multiple charts, with each chart targeting a specific set of audiences and a different broad timetable. Thus, you might have a chart for local/regional audiences; a chart for employees, students, and other internal stakeholders; a chart for partners and potential partners; and a chart for national and international news outlets (the broadest possible audience). Similarly, for events that you know about ahead of time, you may want to further break it up into three sections: a "warm-up" period before an event, such as a groundbreaking or major announcement; a "busy period" during an event; and a "follow-up" period after an event.

Template Chart

This section should lay out how you plan to reach a specific set of stakeholders. Be sure to reiterate and clarify key goals and messages for this section. Here's what a pre-event timetable of deliverables might look like for the opening of our generic example, the JDRC.

Deliverable		*Timing*	*Prime/s*
Pre-opening press engagement	Walk-through and briefing to encourage "this transformational space is coming" articles: • List local and regional new outlets you want to approach. • List other outlets in geographies targeted by your institution for strategic recruitment and giving programs.	[Date]	[Names of responsible parties]
Article on institution's website	• Above-the-fold story on what the new facility will mean for institutional research, student and faculty success, raising the profile of the university, and economic development. • Use social media campaign to amplify message via Facebook, Twitter, etc.	[Date]	[Names of responsible parties]

INDEX